FLYING
SIXTH SENSE

Capital – Quebec City
Largest City – Montréal
Population – 7,390,052
Official Language – French
Motto – *Je me souviens*, ?

Land Data – France was 847,000 km²
Québec, Canada 1,672,329 sq.
km²

Map of Québec.

By
DAVID R MILLS

Table of Contents

Dedication

This book is dedicated to my darling wife, Colette. She has stood by me all these years, raising our two children and working tirelessly through more than a dozen moves. A trained chef, Colette always dreamed of opening a gourmet restaurant, but our frequent relocations made that impossible. Despite these challenges, she has been a loving wife with unwavering support, even teaching me the French I so desperately needed for my work with the police.

I've often heard that behind every successful man is a great woman. In my case, this is undoubtedly true. I am profoundly grateful to have had and still have Colette by my side throughout my life.

Introduction

With a distinguished career spanning 47 years and over 15,000 flight hours, **David R. Mills** has established himself as an exceptional helicopter pilot. Holding the same commercial licenses in both Canada and the United States as airline pilots who navigate Boeing, Airbus, and other major aircraft, his expertise lies in helicopter aviation.

His journey began in the U.S. Army's Warrant Officer Flight Program, where he piloted AH-1G Huey Cobra gunships during the Vietnam War—the world's first helicopter gunship. For his service, he was honored with two Distinguished Flying Cross medals. He is recognized as member #599 of the Vietnam Helicopter Pilot's Association (VHPA) and was also a member of the Distinguished Flying Cross Society.

Following his military service, he returned to Montreal, Québec, and spent two years working on the James Bay Development Project, primarily on the La Grande River dam construction. Seeking new challenges, he transitioned to a role with the Québec Provincial Police, where he served as a helicopter pilot for 11 years. This book delves into his experiences—chasing bank robbers, conducting daring rescues, locating lost individuals in remote areas, and overseeing high-profile events from the air, including parades, races, and even a papal visit.

Through these missions, he developed a unique ability to see with his peripheral vision and learned to trust his instincts—skills that took time and dedication to master, much like learning to speak French.

His career later took him to Fort Worth, Texas, where he and his family spent two years as he trained to become a Test Pilot for Bell Helicopter TEXTRON in preparation for the company's new factory in Mirabel, Québec. A helicopter test pilot is a highly skilled aviation specialist responsible for pushing the boundaries of flight by piloting cutting-edge and modified helicopters to ensure their safety and performance. These professionals operate in rigorous conditions, gathering critical data and working closely with engineers to refine aircraft designs and secure certifications. Their expertise bridges technical precision and innovation, ensuring each helicopter is flight-ready.

Throughout this book, **David R. Mills** shares insights gained from his extensive experience as a combat pilot, police pilot, and test pilot. His knowledge includes crucial lessons that could save the lives of helicopter pilots, many of which stem from his time with Bell Helicopter. Among his proudest achievements is a spotless safety record— never once damaging a helicopter under his command (excluding bullet holes) and never injuring a passenger aboard his aircraft.

Three things one can see and never hear about from any other source.

Three clear, beautiful, perfectly round rainbows inside each other.

A bird (Canada Goose) stalling in a hard-righthand turn, snap rolling into the stalled left wing, and crash landing unhurt into the water, but with severely ruffled feathers.

A bird flying over a thin layer of fog, leaving tracks from the wing tip vortices trailing behind it.

Experience is a hard teacher. It gives the exam first and the lesson afterward.

Learn from the mistakes of others; you won't live long enough to make them all yourself.

The only dumb question is the one you didn't ask.

Nearly all men can stand adversity, but if you want to test a man's character, give him power. (Abraham Lincoln)

Fail to prepare, prepare to fail.

Life is not a rehearsal.

LOW FLIGHT

Oh, I've slipped the surly bonds of Earth,

And hovered out-of-ground effect on semi-rigid blades.

Earthward I've auto'ed and met the rising brush of non-paved terrain,

And done a thousand things you would never care to,

Skidded and dropped and flared. Low in the heat-soaked roar,

Confined there, I've chased the earthbound traffic,

And lost the race to insignificant headwinds;

Forward and up a little in the ground effect,

I've topped the General's hedge with drooping turns,

Where never Skyhawk or even Phantom flew.

Shaking and pulling collective,

I've lumbered the low untrespassed halls of victor airways,

Put out my hand and touched a tree.

Helicopter version of "High Flight" by John Gillespie Magee Jr.

WW 2 Spitfire pilot.

Flying Sixth Sense, the Cover

The Sûreté du Québec was signed into law on 1 February 1870 and predates the federal RCMP by over three years. They are a dedicated body of professional men and women who work to serve and protect the citizens of Québec. They are all well hard-working trained people, and they have many branches that are highly specialized and serve the public well in their specialties. I worked with many of them and was often amazed at their abilities. They enthusiastically embraced the arrival of the helicopter, and many of my successes with the helicopter are a tribute to their work. They are the lowest-paid of the three major police forces in Canada, the RCMP (Royal Canadian Mounted Police), OPP (Ontario Provincial Police), and the QPF or SQ (Québec Provincial Police or Sûreté du Québec). Québec is also the highest-taxed place to live in North America.

In 1947, the USAF and the US Army split and adopted their own wings. US Army wings are shown on the front cover and were awarded to all US Army pilots after 1947.

The US Army 101st Division was established in 1918 and was constituted as Airborne on 16 August

1942. They evolved from horses to parachutes to helicopters.

All US military branches use the DFC. The Distinguished Flying Cross was awarded to Colonel Charles Lindbergh for the first solo crossing of the Atlantic in an aircraft. (His aircraft, named "The Spirit of St Lewis, can be seen in the Smithsonian Museum in Washington DC) Lindbergh received the first presentation of the DFC from President Calvin Coolidge on June 11, 1927. Commander Richard E. Byrd (flight across the Atlantic and to the North Pole) and Emilia Earhart (the first woman to fly across the Atlantic) were also awarded the DFC. There are 2 types of DFC's, one is for heroism, and the other is for extraordinary achievement. A small oak leaf in the center of the ribbon denotes the second award.

This book has one major problem. It is not like a novel that has a flowing story. As years passed, the things I had to do kept adding up, and I saved the lives of a lot of people. But even that got repetitive.

It is a book with hundreds of trees.

I'd get a phone call in the morning. Go somewhere and find a kid lost in the bush.

The next day, it's a photo mission that gets interrupted by a bank robbery. On the weekend, I'd go to a sporting event somewhere for traffic and population control.

Every day was different, and I never knew what I'd do or where to expect to go. Half the Province of Québec was my back yard and I went to every place there was in the province over the years for one reason or another. I saw death in just about all its forms and rescued people from certain death many times.

This book is about the cold hard facts of human life and death and the dedicated police officers that assisted me every time I flew. Fortunately, there were a lot more successes than failures.

I even got kissed by a nun.

Part 1: Flying Sixth Sense

Quaking Aspen or Pando is the largest living organism on earth. There is one tree in an aspen forest in Utah made up of over 47,000 trunks and millions of leaves, all connected through one root system.

I was a miniature of the aspen grove.

The earth where it grew for me was the Québec Provincial Police or "Sûreté du Québec".

The helicopter and I were the roots that connected each day and event.

Each day and event were like one of the trees in the forest of events that occur in this book.

Planting

The Tailor

I closed the garage door after parking the car and climbed the steps into my apartment.

My beautiful wife greeted me with a frown and a light kiss on the cheek and "You stink" instead of a big hug and a kiss.

I knew that I smelled of Jet B, turbine engine fuel, and oil and always had a shower when I got home from my lousy mechanic's job.

Right away, I asked her what was wrong. She told me that my tailor had called.

That was a big surprise. She knew that I was not the suit-and-tie type of man. A uniform, yes. I wore one when I was in the United States Army as a Warrant Officer, CW2. (Chief Warrant Officer 2) I was out of the US Army and back from South Vietnam for two and a half years now. No more uniforms, and I know that I hadn't ordered a suit.

I was wrong, it was ordered for me.

She gave me the phone number to call a tailor on the west island of Montreal.

When I called, I gave my name and the person asked me when I would be able to come to his tailor shop to get fitted for my uniform. I asked for who, and he replied for the Québec Air Service. I told him I would go there tomorrow afternoon.

I gently hung up the phone, and with a huge smile on my face, I bounced off the ceiling, grabbed my astonished wife, and happily spun her around in circles.

"I got it, I got it!" I cried. My eyes were wet, I was in the throes of ecstasy. I was so happy for both of us. Our lives would change. I would be back in a cockpit. I would be with my wife in a job based in a city, not in a tent in the bush. A future with a life for us and a family.

I had worked to be a pilot in the US Army and spent one year in a war zone, later labeled as a helicopter war. The greatest percentage of any MOS (Military Operational Speciality) killed in the Vietnam War were helicopter pilots. The most dangerous job in Vietnam was as an OH-6 scout pilot. The Vietnam Helicopter Pilot's Association (I am member number 599) states that there were over 4,800 helicopter pilots and crew members killed in the Vietnam War. Their names are engraved on the "The Wall", Vietnam Memorial on the Mall in Washington, DC. Then, I spent 2 years in Canada as a "bush pilot."

No wonder I was so deliriously happy. I was looking at a normal life, which in Canada is only about 1% of civilian helicopter pilots.

I explained that the tailor was commissioned by the Government of Québec, Québec Air Service, to make the uniforms that I would need. She knew that I had gone to Québec City to apply for the job and passed an interview. I had been trying to get that job for over three months now. I held her close. I wept with joy.

I got a letter from The Québec Air Service telling me I was hired 2 days later.

That I even knew the job existed was because of my wife.

When I came back from South Vietnam I got out of the US Army and came back to Canada. I had to get my Canadian helicopter pilot's license and find a job.

I had the job before I got my license.

I'd answered an ad in the Montreal newspaper. I called and got the owner; I gave him a brief summary of my experience, and he told me that I was hired if I could pass a flight check. He told me to get to his office in Carp Airport, near Ottawa, the next day so he could give me a flight check.

I told the boss/owner of the company that I couldn't do that until Monday because the Minister of Transport (MOT) was going to issue my license to me on Monday. It was only Thursday. My resumé mentioned my US Army experience and my year in South Vietnam flying AH-1G Huey Cobra gunships in that war.

In Canada, this was a huge experience level over any commercially trained helicopter pilot in Canada. Canadian military pilots stayed in the military. They were all university-trained in the military and then trained as pilots. They almost all stayed there for 20 years. The civilian jobs were too basic, and the pay levels were very low for pilots starting in civilian companies. The helicopters were at a basic level, far below the turbine-powered machines used by the military. For a pilot to evolve in a commercial company would take years of bush, long line, and offshore experience that the military did not offer its pilots.

I picked up my Canadian Commercial Helicopter Pilot's License on Monday morning, and then on to the Carp, Ontario airport, and met with the owner, Gary. He told me that I would have to go to Chibougamau, Québec, and get a check ride from his chief pilot. He had asked me to come and see him just so he could meet me and make his own judgment if he was to hire me. I also left him with a copy of my Pilot's License and my resumé.

The condition was simple. If I passed the check ride, I would be hired.

My problem was that if I didn't pass the check ride, all expenses were to be paid by me, bus, motel, and meals.

As I was starting to leave the office, the phone rang and the secretary called to Gary that the person on the phone spoke French. Before I could get out the door, Gary called to me and said, "Dave, you're from Québec. You talk to this guy."

My pidgin French and his pidgin English, he got his message across. I got a phone number, gave it to Gary, and let him handle it.

I took a bus to the northern town of Chibougamau, 400 miles north of Montreal, and was to go to Motel Alouette in Chibougamau. There I was to wait to be picked up the next morning.

The next morning was typical of a day in early fall in northern Québec, cloudy, with gusty winds and snow showers.

I large red helicopter clattered to the ground in the parking lot of the Alouette Motel. I had never seen that type of helicopter. I had a small bag with me in case of an overnight stay.

The pilot motioned to me from the clattering mass to go to the left side of the helicopter, where I was able to climb up to the cockpit. After climbing up the left side and into the cockpit, the pilot motioned for me to put on the seatbelt and then put on the headset hanging above and between our heads. This I did, and the noisy clattering of the machine was dramatically reduced. The pilot pointed to a button on the floor beside my right foot. I pushed it and the intercom was live. He pointed to a metal rod clipped to the floor on the left side of my left foot. He told me to stick it in the hole between my feet and connect the cannon plug on it to that on the floor. Good, now I could talk to him and control the helicopter with the cyclic stick and the collective on the left side of my seat. I very lightly put my feet on the tail rotor control pedals.

I asked his name, he told me, Jack. I asked what kind of helicopter it was. He told me it was a Sikorsky S-55. (H-19 Chickasaw) I had never seen one or even heard of it.

We were sitting on top of a big Pratt & Whitney R-1340 Wasp piston engine. The cabin was below and behind us. The 10-passenger piston engine clattering machine could lift 2,000 lbs on its cargo hook.

The door closed, seatbelt on, headset on, cyclic control stick plugged in, the pilot took off and then handed me an 8-mile-to-inch aeronautical map, pointed to a lake on the map, and told me there was a camp on the southwest corner of the lake. He then ordered me to take him there. I was now in control of an old, clattering, vibrating machine that was almost as old as I was.

The weather was cloudy with intermittent snow showers. There was good visibility in between the snow showers but it was windy and turbulent.

I was still over the northern edge of the town of Chibougamau so I had a good departure location to start the navigating. I put my finger on the map where we were and stayed low, below 500 feet of altitude. I was terrain following. Hill to the hill, lake to lake, river or stream to the next terrain feature. After several minutes he told me he was lost, he didn't know where he was. I showed him on the map. We argued. I asked him how he could find the lake. He told me he followed the road. The map was new, the latest edition, but the road was too new to be on the map. He wanted to take control and go east to find the road. We argued, and between the snow showers, I showed him where we were on the map. I told him in 10 minutes, we would be over the lake.

He snarled at me, checked the fuel level, and said OK, you have 10 minutes.

9 minutes later, coming out of a snow squall, I could see the lake ahead. He barked that he didn't recognize the lake.

Our relationship for a job interview wasn't starting off too well. I got to the southwest corner of the lake, rolled left, and pointed down at the camp. He told me to land. The Sikorsky S-55 is a 3-bladed helicopter, and to prevent ground resonance, it has 4 wheels to land on. (A 4-bladed rotor requires a 3-point landing gear.) He pointed to a pad made of pine logs nailed together. It was maybe 20 feet by 20 feet large. It was just barely larger than the wheel footprint of the helicopter. The pad was actually made for a smaller machine.

I had never landed a helicopter with wheels.

I had never landed on a log pad.

Slowly, carefully, I landed.

He shut down the vibrating, clattering machine, and we went to the camp.

After meeting the rest of the gang, the pilot took off to carry survey crews out into the bush. I waited in the camp for the rest of the day.

I was assigned a tent with another pilot.

Bob, the other pilot, lay down on his cot, crossed his arms over his chest, and was asleep within 2 minutes. He snored. Flat on his back, a veritable chainsaw. No effort I made could stop him from snoring. I found something to plug my ears, covered my head, and tried to sleep.

Next morning, take a piss, mosquitoes, breakfast, and mosquitoes, check the helicopter, mosquitoes.

Oh, how I wished for dragonflies. My uncle, a high school principal told me about dragonflies when I was about 10 years old. An adult dragonfly could eat up to 100 mosquitoes a day.

Another pilot, who was the company check pilot, took me out in a Bell 47G4. Another piston engine helicopter. It was on fixed floats. He had me do all the emergency procedures. I told him that I had flown the Bell 47 in the military in basic training.

Start, fly, hydraulic failure, then an autorotation. I had to land on the water on fixed floats. I had never flown a helicopter on floats. I had never done an autorotation on floats to the water.

That's what I had to do.

That afternoon, they took me back to Motel Alouette in Chibougamau.

I paid the bill, and the next morning I took the bus home.

Two days later I got a phone call telling me that I was hired.

I had that job for 2 years, during which Gary made me base manager in Chibougamau and, at first, based a helicopter there with me. It was a Hiller 12E piston engine machine. He sold the Hiller a few weeks after I got there to a company in Vancouver.

Gary's instructions were simple. Fly the Hiller to Ottawa for an inspection and then take it across Canada to the part of the company that was based at the Vancouver International Airport. It was early November.

I called my wife and told her to pack warm winter clothes and drive to the Carp airport near Ottawa and meet me there. Two days later, with the inspection done, we left westbound for Vancouver.

We left in the morning armed with maps, credit cards, and a lunch.

Our first leg was from Carp to Mattawa west along the Ottawa River. The weather was low cloud, light rain that soon turned to fog. I tried following the road but the fog got too thick, and I had to come to a hover over the highway, turn around, and go back. When we got out of the fog again, I turned towards the Ottawa River, and there was a clear tunnel just above the water that I could follow. We made it to Mattawa, where we stayed the night.

The next day, the front had passed, and with two more fuel stops, we made it to Sault St Marie. From there, I filed an international flight plan and arranged to meet the customs officer at the small US Sault St Marie airport in town. The next stop was Marquette, Michigan. We were close but didn't make it. I had to stop in a heavy lake-effect snowstorm. I was slowly following the road and flew over a roadside motel. We stopped there and spent the night with the helicopter parked on the front lawn.

By this time, my wife was getting used to the helicopter and was able to eat a sandwich while flying. Her first time in a helicopter was when we took off from Carp, so she had adapted quite quickly. Good thing, too, because she would do a lot more flying in her life.

The snow had passed, so after dusting off about 6 inches of snow from the machine, we set off again. That day, we could only make it to Grand Rapids, the headwinds were just too strong. I could only do about 80 knots, indicated airspeed or 90 mph, and while following the highway Volkswagens were passing us. We spotted some Pronghorn antelope (speed goats) and chased them for a bit until they broke left. I didn't follow them.

Next came customs at the Canadian border. There was no airport nearby, so I landed the Hiller close to the customs office and walked in to declare our return to Canada. We passed just like a car. From there it was off to my uncle's farm just to the west of Morris, Manitoba. My aunt served us tea and cookies, and we spent about an hour with them.

We had to continue our journey. I tried to start the helicopter but nothing. My uncle and I determined that the voltage regulator had burned out. I called Gary. He told me to get the farmer to jump it and then get to Standard Aero at the Winnipeg International Airport. My uncle connected two 12-volt batteries together to give me 24 volts, and the Hiller started up with no problem.

We left with a big thank you to my uncle, and after take-off, I could see 2 men about 50 yards into the field half a mile away. It was my other uncle at his neighboring farm. I buzzed over them, waiving as I passed. I went a little too low, and the 2 of them buried their faces in the dirt. When he found out it was me, he was pissed at me for years.

We made it to Winnipeg that afternoon, stayed the night, and the next day, I got a courtesy car from Standard Aero. I showed my wife where I used to live as a young teenager and the high school in Silver Heights, where I went until I moved to Québec.

We spent one more night in Winnipeg, and then we were off westbound. Headwinds and cold fronts made for slow going. We almost got to Calgary when we were stopped by freezing fog. I landed in a schoolyard and spent the night in a tiny motel.

The next day, another uncle came and picked us up and took us to Calgary to stay at his place for the night.

We got the helicopter to Calgary, refueled, gave my uncle back his car we had borrowed, and then made it to Banff.

The next day was through the mountains and on to Kamloops. Crossing the continental divide was a tight valley to follow. I ran into an undercast of clouds. I tried to stay above them as long as I could but they were turning into a solid blanket. I finally had to dive under that and stay low along the road.

We finally turned south along the Frazer River valley, and we suddenly got a huge kick in the ass. I had flown over a lumber mill, and the steel cone where they burn the bark and wood chips emitted a powerful, hot vertical column of rushing air. I flew right over it at less than 500 feet. There was no damage, but I sure learned something.

Next stop was Vancouver International Airport.

At the office, they arranged a room for the night and dinner.

The next day, they gave me a course on the Bell 206B and turbine engines. He treated me like I had never flown one before. I told him ¾ of my time was in turbine engine machines 3 or 4 times bigger than the little 206B. My case was rare. Very few Canadian pilots had turbine engine experience at that time.

That training took a couple of days, and then we were on a flight back to Ottawa to pick up our car at Carp.

That was one heck of a trip for man and wife. Across Canada in a piston-engine helicopter that took almost 2 weeks and then several days in Vancouver, BC.

To this day, that could be judged as a unique and very special trip.

Upon my arrival back in Carp, Gary had a Bell 206B for me to take up to Chibougamau. My wife went back to Montreal alone with her memories.

One time, I was working forest fires to the north, and on my way back to Chibougamau in the late afternoon, I was crossing the gravel road that was being built to the north, and I saw an ambulance with lights flashing, pulling a long plume of dust behind it, barrelling north up the road. I turned north.

About 10 minutes later, I saw several trucks and other construction equipment with a group of men standing around a man lying on the ground. I quickly landed well clear of the group, got out, and asked what had happened. They told me the man had been trying to dig out the back wheels of a mired truck when the driver popped the clutch, trying to get the wheels out of the hole he was in. The man beside the wheels wasn't ready for it and got pulled under the 4 wheels on the right rear of the truck. They got him out, but he was in very bad shape, and the supervisor got a message to Chibougamau to send an ambulance.

By chance, I had seen the ambulance, and being curious I turned north to find out why. I arrived unexpectedly. I got out the folding stretcher from the cargo compartment and got some men to put the injured man on the stretcher and get him into the helicopter.

The Bell 206 has a unique, well-designed stretcher. The front portion is opened first by opening the front door, and there is a latch on the left door post that unlocks the whole assembly. The door post and left front door fold forward and have a cloth strap with a metal snap that attaches to a snap receptacle just behind the pitot tube in the nose of the helicopter. The left rear passenger door folds back along the fuselage to permit the entire 6-foot length of the stretcher to be loaded into the helicopter. The portion in the front over the copilot's seat narrows so it can fit up beside the instrument panel. The wide part, which is where the head of the patient would go, has special clips to hold it against the back wall of the rear seat. The 2 clamshell doors can then be closed and latched into place.

I then flew him directly to the hospital in Chibougamau, passing over the still northbound ambulance. I had never been there, but I knew where it was. The Bell 206B had a large red cross painted on each of the aft quarter panels of the fuselage, so when I landed at the hospital, it got their immediate attention. I got out and motioned for 2 men with a stretcher to approach the helicopter, keeping them aware of the whirling blades.

I opened the clamshell doors, and they immediately transferred the man from my stretcher to the hospital stretcher on wheels and rolled him into the hospital.

I heard the next day that the man had died from his injuries.

I did not send the bill to the hospital but just added the 5 or 10 minutes of extra flying time to the firefighting flight report. Both bills would have eventually gone to the same place anyway, the Government of Québec, just different budgets.

But the word was out quite quickly that there was an air ambulance available in Chibougamau.

A couple of months later, I got a call from the Director General of the hospital. There was a strike at the hospital and nobody was permitted in or out. He told me there were nurses who had not been home for 4 days and were burning out. He asked if I could bring some new staff in. I told him I didn't want to be a strike-breaker and I would only move essential people in or out. I didn't want some pissed-off striker to damage my machines or complain to my boss. I explained this to the DG. He agreed. He went to the strikers, and they agreed, but only the helicopter with the red crosses on it. We arranged to have 11 nurses transferred in and 11 out.

The nurses assembled at our hangar, and I took my mechanic in with the first load to supervise the loading and unloading of the nurses. I would handle the hanger end. The nurses were able to bring in spare clothing and other essentials for a long stay. I never sent them a bill for the 20-minute flight time. I logged it as a pilot proficiency flight. The strike ended several days later.

Both occasions made me feel good. After flying a Cobra gunship in Vietnam, I was able to help someone instead of killing them.

The activity at the Chibougamau base a few months later was down to near zero, so the boss decided to pull me out and move me back to Montreal.

I was gone so often from Chibougamau and then Montreal to Northern Québec to the James Bay La Grande dam project that my wife gave me an ultimatum, her or flying. I had been gone for a year to South Vietnam and was now leaving her for weeks at a time.

Shortly after we moved back to Montreal, I quit flying. I got a job at Rolls Royce in Dorval, on the west island of Montreal. I was given a job in the turbine engine test bed, repairing and running Nene 10 jet engines used on the T-33 fighter jet and Spey engines used in commercial jets but also on gas pipelines in Canada.

I hated it. Always went home smelling Jet B turbine fuel and engine oil.

I often took time to go to all the helicopter companies that had offices in the Montreal area, Dorval, Cartierville, and St Hubert airports.

I was quite familiar with Cartierville Airport. I used to work there at Canadair Limited. I worked my way up to the flight test department by constantly bugging the supervisor to have me transferred there. I was working with Junior, the lead hand in the flight test. (Junior had a job for life decreed by the Canadair president after Junior's father was sucked into a jet engine intake of an F-86 fighter and killed.)

I worked with Junior where he would get in the F-5 fighter jet cockpit after we had tied the aircraft down. He would start the engines and run them up to full power and also test the afterburners. I would stay outside, plugged into the external intercom to make sure nothing went wrong. We tied the aircraft down in front of a large jet blast deflector. It was winter, and I would sometimes throw chunks of snow and ice behind the afterburners. It was fun to see how far the jet blast would blow the chunks of ice & snow. It was really cool to see them fly several hundred yards or meters onto the long field at the airport edge.

The F-5 had a cool reception for sales to start until the Russians got a hold of a few of them and tested them against the MIG-21. The F-5 would always win in a dogfight with the MIG-21 and could go faster. The F-5 had a unique feature called area rule. It was skinny at the waist where the wings were attached and, therefore, had a lot less airframe drag, which gave it excellent agility in a dogfight. The F-5 was small and cheap to operate, so sales soared after that knowledge from the Russians got out because the MIG-21 was the primary fighter exported by the Russians. (Over 2,600 F-5's were built)

I worked at Canadair for a year and a day. I quit them to join the US Army.

After I quit bush flying, when I was working with Rolls Royce, I went to Cartierville airport and saw a Hughes 269 piston engine helicopter. I knew them well; they were one of two kinds of helicopters used for initial pilot training in Fort Wolters, Texas, where all primary helicopter flight training was done for the US Army Warrant Officer flight training.

There were two stagefields in Fort Wolters, one with the Hughes 269 and the other trained in the Hiller 12E. I was trained in the Hiller.

The manager explained that they were police officers trained as helicopter pilots. He then went on to tell me that the Québec Air Service in Québec City had purchased two Bell 206B helicopters for use by the Québec Provincial Police. He said they had tried police officers as pilots but were convinced by the Québec Air Service that they should get professional pilots. They did exactly that.

The next week, I took a day off work at Rolls Royce and went to Québec City with my resume freshly written.

I met with the chief helicopter pilot. The Air Service only had 1 helicopter and 2 pilots. They were to hire four more pilots for the police.

I gave him my resume.

One month later, I went back to see him. He was surprised. I told him that I didn't want him to forget me.

A month later, I went back again. He told me over 30 pilots had sent resumes for the jobs. I would be sent a letter if I was chosen for an interview. Three weeks later, I got the letter, and about a dozen were chosen for an interview.

I went into a conference room and sat with my back to the entrance door, facing the bright south-facing window. There were 12 people at the table.

They were all French and could also speak English. I could not speak French.

They asked about my experiences. I told them my father was in the military police as an officer, so I knew about that. I had flown for two years in Canada, mostly in Québec, so I knew about that, plus my US Army time and South Vietnam. I had flown the Bell UH-1H and the AH-1G Huey Cobra, both much bigger than the Bell 206B they had purchased, but I had 2 years of experience flying the Bell 206B in Québec.

Near the end of the interview, I pounded my fist on the huge table and emphatically told them that this was the job for me, ex-military, war veteran, and bush experience.

That afternoon, the chief pilot took me for a test flight. I had not flown in 5 months. He handed me a map, 8 miles to the inch, a Canadian Aeronautical Map, and pointed to a spot about 50 miles north in the middle of nowhere up in the Jacques Cartier Park. I flew to the location, proving that I could navigate, and then he told me to go back to the hangar.

I took the map and got the frequencies for the Québec VOR (visual omni range) and the ADF (aircraft direction finder) radio beacons. I dialed the frequencies into the VOR and the ADF, pulled the speed back to 60 knots (best climb speed for the 206B) and 85 torque on the engine (maximum continuous power) until I could see Québec City past the hills. I climbed the helicopter a bit more to 6,000 feet, obeying the NEODD, SWEVEN rule. (North to East, 000 to 179 degrees magnetic odd altitude and South to West, 180 to 359 degrees magnetic for even altitudes.) The VOR and ADF showed about 215 degrees magnetic.

After landing back in Québec, he told me that I would receive a letter informing me of their decision.

During the time I was waiting we went to go up to a cottage that we had bought prior to my going to Vietnam. This time, my wife and I went up with my parents.

One weekend, when I was chopping some kindling from some pieces of pine, I was swinging the hatchet, and my father called my name. I was momentarily distracted, and I caught the thumb of my left hand with the tip of the hatchet. I cut about 3/4 of my left thumbprint into a small thumb-sized steak. The flap of skin was still attached by only a small bit of skin on the tip of my thumb. I held up the bleeding hand and laughed.

Ever since I had come back from Vietnam I was stressed and nervous that something catastrophic would happen to me. While in Vietnam I had heard horror stories of guys who had come home from the war without so much as a scratch on them during their year in Vietnam and after arrival back in the "world" they would get in car accidents, fall, and break some bones, or otherwise get badly hurt shortly after their return from the war.

During my year in Vietnam, I had never so much as scratched myself and was stressed; nervous as a cat, and fearfully super careful about everything I did. I held up my thumb like a trophy, blood dripping steadily and copiously, and went into the cottage, dragging a trail of fat blood drops behind me. I grabbed a tissue, wrapped it around my thumb, got a roll of black electrician's tape, and firmly wrapped my thumb to stop the bleeding.

My wife was adamant that I go to the nearest hospital and get some stitches.

I laughed and grabbed a beer, celebrating the shedding of my stress and tension of waiting for something really bad happening to me. Go to the hospital? Ha! Have another beer. To this day, I have the scarred bump on the thumb of my left hand.

I heard nothing after the flight until the tailor called 6 weeks later, and then 2 days later, I got the letter to make the news official.

Several months later, after I was hired, the chief pilot told me that I was the only pilot he tested who used the navigational aids, the VOR and ADF, to get back to the airport. All the other pilots came back using the map. They were bush pilots, not used to flying where there were navigational aids for the pilots.

Again, my military experience made a difference, plus, I was 13 years younger than the next youngest pilot hired.

Québec Provincial Police

I started on 7 January 1974. I met the other three pilots. Ed was one of the 3, and we worked the same schedule over the years. Ed was perfectly bilingual, so we got along quite well.

The next day, we both took off at about the same time. I went for a patrol and familiarization with the Québec City area. Ed flew to Montreal to set up the police presence in that district.

We were two pilots for Québec City, and the Air Service had set up a 4-day on, 4-day off work schedule.

Snowmobile bank robbery:

I was hired to fly for the Québec Provincial Police on 7 Jan. On 17 January, 10 days after I started, I was at home & got a phone call to action at about 9 am. A bank robbery had just occurred in a bank in Inverness, a small town between Québec City & Thetford Mines. I flew down with my police observer & picked up 2 SQ (Sûreté du Québec) police officers who had come from Thetford Mines. After I took off, the observer with me was talking to other SQ officers in cars on the ground. Witnesses had seen a snowmobile at high speed racing down the river westbound from the Town of Inverness. Inverness is out in the sticks. Everyone had a snowmobile. There were tracks everywhere. I picked up on the info & stayed high. There was not a straight track anywhere except one going west. I dropped down to about 20 feet & slowly started following a snowmobile track that was obviously running at high speed. I noticed that every time there was a bump, the tracks jumped the bump. The rubber track of the snowmobile went airborne a few inches where there was no track in the snow, showing that the snowmobile was moving at high speed. Curious. I followed the track for a few miles & noticed no less speed. The river was a chain of very small, narrow lakes. The track stayed straight down the middle of the river/lake chain. Several minutes & far ahead, I saw a snowmobile stopped in the middle of the river. The snowmobile had bogged down in the slush of the river. A river flows down. The ice prevented the water from emptying fast enough to prevent the water from flowing over the ice, unseen, under the snow. Travelling downstream they had encountered the water that was over the ice and under the snow, causing a layer of slush that trapped the snowmobile.

There were the slush-filled tracks of 2 people leaving the snowmobile towards a small forest of trees, mostly maple trees. With the leaves gone, there was no cover. I followed the water-filled tracks to the edge of the river and then into the small forest. I did not see any tracks out of the trees. I went back to the snowmobile & at a very slow 10 feet, I followed the tracks back to the trees. The tracks wandered around the trees & were no more. I saw a bump in the snow. My SQ observer was just along for the ride, and he didn't really know what I was doing. It was his first time in a helicopter. I told the SQ police observer that we had them. I went into a high hover at the edge of the tree line & hit the siren. I then took the loudspeaker microphone off the hook & handed it to the cop. Excitedly he grabbed the microphone & started yelling into the clip on the back of the microphone. I took his hand & rotated his hand 180 degrees so that he was yelling into the microphone.

The 2 bank robbers slowly shook off the snow & trudged into the open beside the trees. They were armed robbers. The cop had pulled his revolver from his holster & pointed it at the 2 suspects. I pushed a little right pedal while in the hover to put them on the left side of the helicopter. That put the cop between me & the two armed robbers. The cop, who still had his revolver pointed at the 2 suspects, was pointing it at them through the windshield. I reached across in front of him & opened the small sliding window on the co-pilot side door of the Bell 206B & then grabbed his gun-toting hand & pushed it to the open window. The shoot is OK, but not through my windshield.

I landed & the 3 cops exploded from my Bell 206B.

The 2 suspects were thrown down into the snow & handcuffed. One of the officers went to the tree & after around thrashing in the snow for a few minutes, he came up with a pillowcase full of cash &

a starter pistol. (Starter pistols make a noise like a gun but only shoot blanks)

That cop brought the pillowcase full of cash & gave it to me. I opened the pillowcase. Bundles of cash. 100s, 50s, lots of 20s, 10s, & 5s. Well over $11,000 in cash. Not much, for which the bank robbers would eventually receive 5 years each in prison.

3 cops, 2 bad guys, plus me. Not enough seat belts. 1 cop & 1 bad guy were left behind. As I took off, I saw a farmhouse about a mile ahead. I started to make an approach to the large plowed parking lot in front of the barn. I heard loud, squawking noises from the back seat. The translation was that his uncle lived there. Go someplace else. OK. I turned east across the river & landed in a wide snow-cleared driveway beside another house. 1 cop & 1 bad guy got out. We went back to the field & picked up the other cop & bad guy #2 & then went back to the house. I shut down the helicopter & went into the house for a cup of coffee. Thank you, it was a stimulant, but it was also a stress reliever. 2 cops arrived from Thetford Mines. The 2 bad guys were loaded into the cop cars & with great reluctance & a big smile, I gave the cop the cash-filled pillowcase. My cop observer and I then went back to Québec City.

Upon arrival, back at the Québec Air Service in Québec City, the dispatcher told me to go and see the

Director General of the Air Service. The DG told me that he had received a congratulatorily call from the Minister of Justice, who was ecstatic. It had taken only 10 days to get great recognition for the purchase of the two helicopters. It was him, the Minister of Justice, who had pushed the budget through parliament for the purchase of the two helicopters.

CHP: (California Highway Patrol)

The next day, I was sent up to Lake St Jean. There were to be snowmobile races and the cops wanted me up there. Again, the language problem had to be worked on. I was starting to learn French, but French is a very difficult language to learn. Too many conjugations of all the verbs, and everything is masculine or feminine and quite simply has to be memorized.

The problem was getting there. I could check the weather in Québec and then Roberval and Chicoutimi in the Lac St-Jean area, but the mountains in between were the problem, for there was no weather reporting station for the 120+ miles to get through them. I had to follow the road just to keep a positive reference point for the trip north. I often had to let the road climb up into the clouds and then try to find a valley to get around the areas of low clouds. The other problem was fuel, there was no place that I could refuel in the mountains. Over the years there were several times that I had to hover just over the highway for several miles in the cloud.

Wires were a killer. I had to learn where they crossed the road when hovering up in the clouds. The big transmission lines were on the map, but they crossed the road many times along the highway.

Getting to the destination was often the biggest challenge.

The police had several cars in the area, and I was there for traffic control and to serve as an air ambulance in case of an accident of the snowmobile racers.

We were watching the races from above and some car drivers wanted to race as well. The traffic was heavy, so many were passing other cars illegally. The speeders were passing across double lines on curves and intersections.

We set up a couple of marked police cars and spent a couple of hours giving tickets to those impatient drivers who broke the traffic laws and caused the accidents.

One of the problems we had was identifying the police cars my observer wanted to talk to.

We put up with this problem for a couple of years. I often watched a TV program called CHP. California Highway Patrol. I noticed that all the police cars had their car number painted on the roof of the vehicle in about 2-foot-high letters.

I mentioned this to the Director General of the Québec City district. It took a year or so, but the Québec Provincial Police finally did the same. I like to think that I may have helped them to make that improvement. After that, my observer and I could easily identify any police car that we wanted to contact.

Québec Winter Carnival:

The Québec Winter Carnival is held at the end of February and attracts several hundred thousand people. There are ice castles and anything else imaginable that could be made or done with ice and snow, including boat races across the ice-filled St Lawrence River just below the fortress walls. One lower town street had ice figures the length of the street, and "Caribou," made with red wine and vodka, was the main fuel carried in a hollow cane that held about a liter or just under a quart of this potent mixture. Many ice sculptures were very artfully done. Almost all had a little kiosk set up to sell souvenirs, plus the inevitable "Caribou". When you are outside in the freezing cold, you don't get drunk, but if you walk into a hotel or a bar, the heat will hit you, and you go from OK to drunk in just a few minutes. I know, I did it once. It hits like a truck.

There are also 2 parades, one each Saturday night. The first one is in Lower Town, and the second, a week later, is in Upper Town. From the University of Laval down what is now Rene Levesque Street to the fortress walls in front of the parliament building.

We discussed the huge crowd covering the upper town route and the parliament area. I told him I guessed at about 500,00 people. My observer radioed that comment down to the other vehicles on the ground. The next day, that number was in the newspaper. Back then, police radios were not encrypted and that information had obviously been heard on a scanner by the media.

After the parade passes and ends in front of the parliament building, the fireworks start.

Mounted on the helicopter was an external "Night Sun" twenty million candlepower light. At over a thousand feet high, a person on the ground could easily have enough light to read a newspaper. (To turn on the "Nightsun," I had to turn off the heater fan. The helicopter generator was simply not powerful enough to supply electric power to both of them. The "Nightsun" demanded so much power from the generator it showed high in the yellow range on the Generator Loadmeter when the light was on).

The first year I was there, I flew over both parades. They were fun to watch, and I had a QPF officer and two Québec City police officers with portable radios for their police frequencies. They guided police cars from the back of the parade to the front as the parade progressed. The back of the parade also pulled thousands of spectators to the huge ice castle in front of the parliament buildings, where the fireworks would sparkle and light up the night sky. One-way streets didn't matter. We even found some parking spots for some of the cars. Crowd control from the air was very effective.

There was one unused seat belt between the two cops in the back seat. I brought my wife on several occasions over the years. Both my children had a helicopter ride before they were born.

THE LETTER!!!! I PANICKED!!!!

I received a letter from the Government of Québec.

They put me on a probationary term of one year to learn French.

I was now in a helicopter pilot job that I really liked. Instead of being gone for weeks at a time, I would be gone from home only 2 or 3 nights a month.

Learning to read, write, and speak French in one year was a daunting task. It would take about 600 hours of classroom instruction to learn French or any one of the many European languages.

I went to every school and university that I could find in Québec City. None taught elementary French. I found a few that taught advanced French for aspiring writers, journalists, and teachers but nothing for an entry-level student. We were in Québec City. 99% of everything was French.

I went to the chief pilot and showed him the letter. He told me he could do nothing. The police were my boss.

The headquarters for the Québec Provincial Police (Sûreté du Québec) was in an old hospital in downtown Québec City. I wore my uniform, and when I got there, I was told to go down to the basement. I was looking for the Emergency Squad (Unité d'Urgence). I was directed to the Captain's office.

Entering the office, I encountered his secretary. I asked to speak to the Captain. She answered me in English and led me into his office.

On his desk, I immediately noticed a wooden plaque with "Captain" with his name engraved beside it. I noticed that Captain was spelled in English, not in French. Capitaine.

He knew who I was right away, having heard me speaking English and seeing me in my uniform. He invited me to sit down, and we chatted for a while. He was happy with my work so far, and he told me that, like me, he was English, and because he could not write French well enough, he had to have a bilingual secretary. He went on to tell me he was limited to being a Captain for that reason and could not advance to the rank of Inspector.

I showed him the letter. He read it, looked up at me, and smiled. I was surprised at his reaction. He understood my problem and told me he would help me. He would talk to all the SQ officers under his command and find out how many could speak a little English. He would then use these cops as observers for me in the helicopter, but I needed to do my part. That statement gave me a huge measure of relief. I told him my wife was French and I intended to subscribe to the Québec City daily newspaper. We talked for a little longer and we were able to better understand each other and for me to know what he wanted and for him to learn what I was able to do with the helicopter. He had never had a helicopter under his command before and knew very little of our capabilities. It was great, we both learned a lot about each other. He was easygoing and treated me as an equal. He had not seen my resumé, so I told him about my previous experiences. He was delighted and knew all about the bank robbery and the apprehension of the two bank robbers. I was highly motivated and delighted with his parting words. "Keep up the good work, Dave."

Now, I really had to get to work to learn French. I would sit at the kitchen table every day and read the newspaper out loud. My wife would correct my pronunciation and translate any word I did not understand.

It was fantastic. I was married to a French dictionary.

There were a few light moments doing this. One was "nid de poule." I pronounced "nid" like Sid or bid. She didn't understand what I was saying and looked at the newspaper and laughed.

She told me a "nid de poule" was a pothole in the road. It was a written report about a road that had many potholes and had caused damage to several cars.

Nid in French was pronounced like knee, the "d" was silent. So "knee de pool" was the correct pronunciation. The chicken nest was the direct translation of "nid de poule." So, the road in question was full of chicken nests.

That reminded me of another time when we had fun with French pronunciation. I was taking my wife, then girlfriend, out on our second date. We went to park in a very large parking area that used to be a parking area for the "Expo 67" held in Montreal.

Each section was identified by an animal. Giraffe, elephant, horse, etc. She told me in French "Nous stationnons au phoque". I was shocked! Our second date, and you want to park and fuck? "Non, non," she cried, and pointed at the sign, "un phoque, un phoque". The sign showed a picture of a seal. We would park in the section of the seal. An easy way to remember where your car was in the huge parking area. How huge? The area later became an STOL airport. (Short Take Off and Landing).

A couple of years later, when we were married, I was on a course in Fort Eustis, Virginia. I met 2 lieutenants on the same course, and we decided to get together with our wives and go for a nice meal at the Newport News Naval Officer's Club.

The six of us got together for a drink before going to the Officer's Club. We told them about the parking lot confusion.

When we got into the club, one of the lieutenants said, "This is a sealing nice club". Well, that started it. All during the meal, it was pass the sealing salt, order another sealing bottle of wine, or one lieutenant said to his wife I want to seal you tonight.

No person listening could understand what we were all laughing about, but we all spent the evening in a very elegant Naval officer's club, ate a delicious meal, and all had a seal of a great time.

The observers now assigned to me could speak a little English, and from that point on, my observers had enough English to understand me, and that helped me learn a lot. They would also translate the radio chatter.

In the weeks that followed, I continued doing a lot of patrols around the Québec City area, learning the territory, the way the police operated, their radio frequencies, and, of course, getting to know my observers and how they wanted to operate. This was a huge learning experience for me. Everything was new. Everything was different. I loved it. When the phone rang, it was always a new adventure. I'm sure it is the same for all police pilots around the world.

My case was rare, I was not a police officer. Almost everywhere in the world the pilots are police officers first, and pilots second. What was better? I don't know, but my pilot skills had to be very good, if for no other reason than the size of the area where I worked. The Province of Québec. It is the largest province in Canada. Québec is 2.2 times bigger than Texas. Québec is larger than Alaska. All of France would fit in the uninhabited northern part of Québec. Two helicopters covered this vast area.

At home, my wife, my French teacher, started correcting every word or phrase that I spoke. If I pronounced a word wrong, had the wrong tense, or got the masculine or feminine wrong, she would correct the word, and I would go back to the word, correcting it in my speech as I continued to talk. I learned my French by memorizing it, just like a young child. I never wrote French, don't to this day, but I can talk in French about almost any subject. As I was learning my French, my wife was still learning her English, and in the beginning, one of the words she would mix up was chicken and kitchen. She would say, "I'm going to cook a kitchen in the chicken." She would wix her murds just like me sometimes.

From 4 to 7:

The shift changes of 4 days on and 4 days off just didn't work. One of us would go somewhere for the weekend, and the change would occur on Friday or Saturday after the machine had arrived for a job somewhere. We would commute back to Québec City using a relay of police cars. One would meet halfway to the destination, we would switch cars, and the police car would return to departure point with the other pilot.

The other problem was that we both lived in Montreal, so we each had to commute back and forth every 4 days. This was in the middle of winter, and snowstorms were frequent. I had left Montreal, heading towards Québec City with snow falling, when I saw a pile-up of cars under an overpass. It had happened seconds before I got there. I had seen the lights crossing and flashing brake lights a couple of hundred yards ahead of me. I stopped and got out to help and went to one car that was badly smashed. The lone person inside was obviously dead, crushed, and cut up. I told those around the accident that I was going to call the cops. I weaved through the mess and to the on-ramp, (the off-ramp was behind me now plugging up with traffic) up to the road crossing over the highway. I then turned to the nearest house and told them to call the cops and send an ambulance; at least one person was dead. That was the first dead body I'd seen since I got back from Vietnam, but it was not to be the last, not by a long shot. I went back to the accident site, and when I saw the cop car arriving, I left and continued my trip to Québec City.

I called the Emergency Squad Captain and suggested a "7 on 7 off" schedule. He told me no problem, just as long as he had a pilot for the machine. We decided on a Monday night changeover that would work well for long weekends as well. I called Ed in Montreal and they happily adopted the same schedule as us. All the rest of the years I worked for them, it was a 7 on 7 off schedule.

That worked really great when I took 7 vacation days, I had 21 days off.

Flood:

March 5th dawned clear, warm, and sunny. My observer turned out to be the Corporal of the Emergency Squad. He wanted to spend some time with the helicopter. He had been sending his men to be my observers and wanted to know what they were doing with the helicopter. That day, he learned fast, the hard way.

A few minutes after we took off for a local patrol, we got a call that sent us to St Clothilde, a small town on the side of a river south of Three Rivers, which is between Montreal & Québec City, along the St Lawrence River.

The rapid snow melted and the still frozen ground was causing flooding. The river was well over its banks and into several houses. When we arrived over the village, we could see the river had widened by several hundred yards. We located the police cars that were on the high banks of the north side of the river. The water was over the bridge that crossed the river and many houses on the south shore were in several feet of water.

A woman and her 3-month-old child were reported isolated in one house with no power and no rescue access due to the high flood waters. The house sat alone, surrounded by trees. The husband had tried to rescue his family and was now in the hospital with injuries sustained during his rescue attempt.

The house had a chimney at one end and a high front porch at the other end.

The chimney was a major danger for my tail rotor of the helicopter, so I had to angle the landing on the roof of the house to 40 to 45 degrees. The surrounding trees overhung the roof and I had to chop the leaves at the overhanging tips of the spreading branches to be able to lower the helicopter to the peak of the roof. I could not put the full weight of the helicopter on the roof, it was too heavy, plus my skids of the machine were at a 40+ degree angle, which meant that I would have to fly the helicopter even with my skids touching the roof. The snapping of the leaves being chopped off by the rotor blades, I would later find, had stained the rotor tips to a bright green. There was no damage to the blades for I was only chopping the tips of the tiny branches and their leaves overhanging the roof of the house. I lightly settled the helicopter onto the roof.

The corporal slowly climbed out of the left front seat. He had to transfer his weight to the roof slowly to not upset my balance of the machine on the roof. I had explained that a sudden weight transfer would be dangerous. He had to go back to the cargo compartment to get a rope I kept there. He got that, also slowly. He used the rope to climb down to the porch to get the woman and her baby. She had to pull a small table out to the porch and put a chair on top to permit the corporal to climb down. He first carried the baby up and brought it to the left door of the helicopter. We secured the baby in the front seat with the seat belt. When he was securing the baby, the corporal looked across at my right foot. My left foot was holding firm on the left pedal, which required force since it was the power pedal. My right foot was nervously and rapidly lightly tapping the right pedal. My knuckles were white on the controls. The pucker factor was extremely high. He looked up into my eyes in understanding and closed the door.

He brought the mother to the left rear passenger door, and she, as instructed, slowly got into the back seat. The corporal followed, and with seat belts secured and the door shut, I slowly lifted off the roof. The snapping off of a few more leaves were unnerving, but then I was clear. The stress of those several minutes had left me tensed up like a tightly coiled spring. I flew across the river, where some police cars were parked. After landing, I let go of the controls and totally relaxed my body, trying to relieve the massive stress buildup of the last 20-some minutes. I realized that I could hear music. Prior to arrival, I had tuned into a Montreal AM radio station on my ADF. My total concentration on my flying had blocked the music from my senses. In the few seconds that I had relaxed and started to shed the stress, my mind suddenly heard the music again.

The corporal helped the woman and her baby get clear and then he spent a few minutes talking to the other cops.

After getting back in my machine, he told me there were over 22 more people in 3 different houses. Two houses had people that were all OK. They were dry and just stranded without heat or power. The third house was the big problem. They were also dry, but the lower part of the house was underwater. It would have been easier to land on the roof of the house but there was a tall TV antenna in the center with guy wires to each of the four corners.

The house had a barn or large garage beside it. In between were parked a 5-ton truck with its nose to the barn and a pickup beside it with a small boat in the back.

We were lucky; it was perfect for the rescue. The corporal would have to get wet.

I carefully landed the helicopter on the top of the box of the 5-ton truck. Here, I had to be very careful because with me landing very lightly on the extreme back of the truck, my rotor blades were less than 2 feet from the barn roof. I again could not put the full weight of the machine on the box of the truck. The box would collapse with the full weight of the helicopter on it. That didn't count the weight of the people who had to walk on the top of the truck's box and then get into the helicopter. I had to pull more power to compensate for their weight yet remain in place on the roof of the truck. Rigidly unmoving was harder than any action possible.

The corporal again slowly got out of the helicopter and onto the top of the truck's box. The corporal then went to the front of the box, climbed down to the roof of the cab, then down to the hood.

From the truck hood, he jumped across to the pickup cab. The corporal then got down into the box of the pickup, where he untied the boat then added another piece of rope to have enough length to be able to cross the cold water to the house.

The water was snow melt, ice cold.

The side of the house facing the barn had only one window. It was a little bathroom window, and 3 people had to climb out that window to let the corporal help them into the small boat. They crossed to the pickup, then to the truck where the corporal told them to wait. I wondered why.

The corporal pushed back through the freezing waters, dragging the little boat to the house, and the people there helped a 14-year-old boy with a cast on one of his legs climb out the window into the boat.

Now I knew why the others waited.

The corporal brought him to the pickup and then went back for the last 3 people. He got them to the pickup where they all worked together to get the boy with the cast over to the box of the big truck. I took 3 in the helicopter over to the waiting cop cars, then back for the boy with the cast. With his plastered leg, the boy needed the whole back seat. When we got to the other side, the corporal helped the boy to one of the police cars and we went back to get the last three people.

The times I had been waiting for the corporal to get the people into the machine, I called over to the other cops and told them to get the corporal a complete set of dry clothes, including foot-ware. They did. We dropped off the last three people.

I then went to Three Rivers for fuel. I was gone almost an hour, which gave the corporal time to change clothes and warm up.

When I got back and picked up the corporal, we went and landed on a large patch of grass beside the last two houses. This was easy, but the last group of people had an elderly person in a wheelchair, and of course, the wheelchair had to come as well. It took several trips to get the rest to safety.

The corporal was almost blue from the cold and was still shivering, even with the dry clothes. I had kept the helicopter heater at maximum heat since his first jump into the freezing waters.

We were finished, and I needed fuel again. We went back to Three Rivers for fuel. The Government of Québec had a Ministry of Transport enclosure with offices, snowplows, and large storage sheds for sand and salt for the roads in the winter.

The Québec Air Service had put caches of fuel, Jet B, in 45-gallon barrels, (55 gal US) in several of these enclosures throughout the province. With the heater going full blast and the exhausted corporal sound asleep, I landed in Three Rivers. I emptied one barrel of fuel into the tank of the helicopter and then flew back to Québec City with the corporal still asleep.

The corporal learned the hard way what the observers he sent to the helicopter might have to do. He never came with me again.

I flew 6.1 hours that day.

This time it was the Minister of Transport who called the Director of the Air Service with his appreciation of the rescue of so many people from the flood waters in St Clothilde.

Someone had taken some pictures of my helicopter when I was moving the people, and I heard that the pictures were in one of the Montreal newspapers. That was good, cops always like and need good publicity once in a while.

After that, I noticed a subtle change in the way I was treated by the Air Service people and the cops. I got anything I asked for. I had expressed a desire to have fixed floats installed on the helicopter. They were large rubber float bags, and I could land on the water and shut down, then restart to take off. One thing that spiced up the start was the torque induced by the engine power turning the rotor. The torque caused the helicopter to rotate to the right, and I had to hold the full left pedal until the tail rotor RPM (revolutions per minute) was enough to control the spin. The helicopter would rotate one and a half turns before I had sufficient tail rotor RPM to stop the rotation. The anti-torque tail rotor pedals controlled the spin, and any time I would pull power by increasing the rotor blade angle the machine would require more power to the tail rotor by increasing the pitch on the tail rotor blades. I'll explain more about that later.

North of Québec City is the Orsainville prison. There was a prison break, and I was called to check the area. There were no tracks in the snow anywhere near the prison so it must have been a planned break with an accomplice ready to pick the escapee up. He was on the loose for about a week when he was picked up & put back where he belonged.

There was a noticeable decrease in the number of attempted prison breaks. Those inside knew that they had to contend with a helicopter, not just hide in the bush for a few hours to get away.

Tickets:

The cops were starting to like to use the helicopter to give traffic tickets. The Transport department started painting triangles every 500 meters along sections of main highways that attracted speeders. These they would paint on the paved shoulder of the highways for a length of 3 or 4 kilometers, and the police officer in the helicopter would use 3 stopwatches tied together on a board, and the time would give the speed. Fifteen seconds was 120 kilometers an hour. When the speeding vehicle was clocked at over 120 kph, they would be pulled over by the cops waiting along the highway & given their tickets. No tickets were given below 120 kph because the fines above 120 kph were much higher, plus 120 kph is where demerit points would start to be given. The demerit points lasted for 2 years on the license. Over 12 points, and the driver would lose his license.

There were a few tickets that got a lot of points. Passing a stopped school bus with red lights flashing was 9 points. There were times when I would follow school buses, and on occasion, we would catch someone passing a stopped bus with lights flashing.

There was a highway #54 north of Québec City which was a popular place for speeders and illegal passing across double lines. Over the years the police used this stretch of highway as a popular place to give tickets.

Moose:

One day, we were doing just that when we got a call about 4 male moose along the highway in the Jacques Cartier National Park. Moose scared the police. A male moose could weigh between 1200 to 1600 lbs, (540 to 725 kg). A police car had previously hit a moose when coming over a hill, and it took the top of the car completely off. The cop inside ducked and only had several cuts and bruises, but the car was scraped.

I found and then herded the moose several miles to the West.

That highway now has a fence on both sides of it to prevent moose or bears from getting to the highway. That was done during the construction of the 4-lane highway, expanding the former 2-lane highway to 4-lanes.

This was not the only encounter I had with a moose.

A couple of years later, a DC9 airliner at the Québec City airport had to execute a go-around because there was a moose on the runway.

I arrived in, about 20 minutes after the call, and a local city police officer was waiting for me. I called the tower and told them I was going moose hunting and the airport was mine. No other airborne traffic was permitted. We took off and found the moose in minutes. There was a Game Warden officer with a rifle at the west end of runway 240, near the approach end of 060. East of the airport were hundreds of houses. West of the airport were open fields. I herded the moose across the runway, but the warden was busy chatting with some spectators and couldn't shoot. That was good because the warden placed himself at the button of Runway 060. If he took a shot, he would be shooting to the east, towards the houses.

I told the cop beside me to open the sliding window of the door on the Bell 206B and shoot the moose with his revolver.

We could hear the meaty slap as each of his 6 bullets hit the moose. The first 6 bullets didn't even slow the moose at all. The cop reloaded his revolver, and as I herded the moose from about 15 feet (5 meters) away, he put 6 more bullets into the moose. By this time the moose started to slow down. We were getting near the northern boundary of the airport along the fence, and when we reached it, the moose sagged against the boundary fence. The game warden arrived a minute later and finished the moose with a .30 calibre rifle bullet to the brain. (Shooting any animal from the air is expensively illegal in Canada.)

The moose was later picked up and was served in welfare kitchens in the city. An unheard-of treat.

I quickly landed back at the Air Service hanger and told the tower controller the airport was his again. That was the only time I had to take control of the airport from the tower controller, and was the only time I had even heard of any other occasion of that happening anywhere. Airport controllers "own" and usually control airspace 5 miles from the center of the airport and from the ground to 2,500 feet above ground level (AGL), however, that does depend on the size of the airport. They also control all aircraft and vehicle movement on the ground. Higher than 2,500 feet (AGL), pilots still had to get permission, but from a different controller.

For about 15 minutes, control of the Québec City airport was mine.

Minutes later, the DC9 airliner landed safely.

Eastern Québec:

I continued my patrol routine around the district and other cities in the eastern half of the province.

Québec is the largest province in Canada, larger than Alaska and more than double the size of Texas. The police had two helicopters to service this vast area. Almost all my flying was done in the southern part of the province where the people were, but there were odd occasions when I was sent to some of the far reaches of the province.

I was kept busy flying between 35 and 50 hours a month on a 7 on-7 off schedule.

The helicopter needed a 100-hour inspection about every 5 to 6 weeks because the other pilot flew as much as I did.

Professional commercial airline pilots can only fly a maximum of 80 hours a month, and that is with 2 pilots, an autopilot, and a flight director to do most of the work. I flew half that alone, with no autopilot, and not at 30,000 or 40,000 feet for long distances, but short distances and close to the ground. Wires and antennas are a helicopter pilot's greatest enemy.

Blood drive:

Near the end of May, we were tasked to go to Montmagny, a town east of Québec City, downriver on the south shore. I was to go to a Red Cross blood drive and fly the donated blood back to Québec City. The Red Cross asked for the helicopter to quickly get the fresh blood back to a refrigeration unit in the city.

Montmagny was several miles to the east, on the south shore, the highway bridges were to the west of the city, and the Red Cross receiving offices were at the east end of the city. I logical request. The cops agreed, and we took the director of the Québec City district of the Red Cross with us, and on the way there, we switched the police radio to the Montmagny frequency. Right away, we heard cars looking for a Ford Galaxy 500, red and black. I positioned the helicopter high enough to see the road along the south coast of the St Lawrence River and the 4-lane autoroute a few miles inland. As we approached Montmagny, I saw a black & red car crossing an overpass a couple of miles ahead. He pulled into a small hotel/bar/restaurant at an intersection south of the autoroute. The car matched the description of the vehicle the police were looking for.

The red and black Ford pulled into the large hotel parking lot. I followed him & was preparing to land when the driver saw me and floored the car's engine. He spun around and passed directly under me. If he had hit the helicopter, he would have flipped me out of the air. That was a killing move on his part. I was enraged.

He was going to get caught. My observer called in the 2 police cars that had been searching for him. I instantly followed him, staying just above the tops of the trees and the telephone poles. With my siren screaming as loud as my rage, I stayed as low as I could right on his 6 o'clock. Soon, I had to climb a bit so the police cars behind me could see how far behind us they were.

The road made a long curve while going through a wide band of trees. On the curve in the trees, he missed a head-on collision with a dump truck by inches. The Ford went through several curves with the car leaning precariously on the tires. Coming out of the trees, he hit a straight section of the road. The left front tire of the Ford Galaxy 500 blew. It slowed him down, but not much. He was running on the flapping rubber of the tire and then just the rim. He kept the big Ford V-8 floored.

Smoke started leaking from under the hood. He didn't slow. Finally, a cop car had caught up to him and was cutting him off. I landed on the road in front of the Ford and my observer jumped out and streaked to the car, now billowing white smoke from under the hood.

The cop from the police car was out and jerked open the driver's door. He hauled the driver out of the Ford, and my observer got there quick enough to catch the driver & throw him to the ground. The passenger of the Ford was also hauled out and pushed into the back seat of the first police car. The second police car arrived and he helped pile the now handcuffed driver into the first police car and then stayed to call a tow truck for the Ford. The first police car headed back to Montmagny. He went to the Montmagny police post with the driver and passenger from the Ford. That's when the sergeant took over.

My observer got back into the helicopter, and we headed to our rendezvous with the Red Cross. I looked back at the Red Cross director, and he was white as a sheet and wide-eyed, but he didn't get sick.

Later, I landed and shut down the helicopter and went into the school where the Red Cross blood drive was being held.

While the blood was being prepared to be loaded into the helicopter, one of the pretty nurses came and asked if I wanted to give blood.

I told her that federal laws, the CAR's (Canadian Aviation Regulations), would prohibit me from flying for two days after giving blood. So, no, I couldn't, I had to fly the donated blood from the blood drive to Québec City. She told me OK, but I could have some coffee and donuts anyway.

A while later a few hundred pounds of blood were loaded into the helicopter, and we headed back to Québec City.

It was about two months later before I got back to the Montmagny police post. The sergeant in charge called me into his office and told me what happened to the two young men in the Ford.

The passenger could not be charged with anything, so they called his father to come and pick him up.

The driver was a different matter altogether. He'd broken a lot of traffic laws and was surly and arrogant. The sergeant called the father and read him the list of fines he would have to pay, plus the car was disabled and had to be towed.

The young man's father refused to come and get him or pay the fines. The father told him the kid was arrogant, disobedient, and swore at his parents, and the father told the sergeant to keep the kid because he couldn't control him.

The sergeant of the post of Montmagny realized that the kid lacked discipline. Discipline was part of his job and he knew how to do that.

He then did what could never be done today.

First, he gave the kid a cell, that was his room.

He waited until later when he had 2 shifts of police officers under his command in the post at the same time. Their job was to teach the kid discipline. He told them all what had happened and what he wanted, and the rest of the cops later.

The sergeant was to give the janitor a few weeks of paid time off not counted as vacation time. First, he had to spend a day to teach the kid what he had to do.

When the kid arrived, he was surly and arrogant. The cops were going to change that attitude, and they had all the time they wanted.

Every cop was to be called Papa. The kid soon learned that sleeping on a wet bed with wet clothes was extremely unpleasant. The bathrooms had to be spotless every hour. The kid would keep the building and police cars clean, and the cops knew where the muddiest and dustiest roads were. He would shine boots and shoes and make and serve coffee to whoever, whenever they wanted. Cops understood discipline. Coffee cups would often fall off a desk or counter. They had to be replaced and the mess cleaned up. Boots got very muddy and so did the police cars. The back seats in a cop car are all lined in plastic and needed to be washed every day. The floors in the building needed to be cleaned and shined every day. All the cells had to be kept clean. Every cop was Papa, and he learned to say please and thank you a lot. He was fed if the jobs were done right.

There were three shifts of cops each day, and he was lucky if he got 6 hours of sleep a night. Work, serve, and clean, all politely and immediately.

The sergeant said he called the father to come and get his son after 4 weeks.

A few days later, the father called back crying. Thank you for giving me my son back. He helps clean around the house, does what he is told, doesn't swear at his parents, and is polite and helpful.

The sergeant told me he passed the message to all his men.

Job well done.

What the sergeant told me was the only thing that happened in this book that I did not do myself or personally witness.

First search:

The next day, I was called to search for an elderly lady who had gone for a walk up North of Québec City in the cottage area and had not returned. I couldn't find her. That was frustrating.

The next day was a changeover, and the other pilot found her. That bugged me. I should have been able to find the woman myself.

A few days later, I decided to go for a walk in the woods. I decided I would look for anything that was not supposed to be there and movement. A squirrel moving, a bird flitting among the trees, anything that moved.

I was surprised that I really had to search for what was there. The human eye is automatically drawn to movement, but I also tried to see anything that was not supposed to be there: a piece of paper, a discarded soda can, anything. I was soon surprised at how long I had to take to see what my eyes were looking at. I realized I had to train my brain to see what my eyes could see. That took time and a lot of walks in the bush.

I did that when I was flying as well, and that also developed a kind of 6th sense that drew my eyes to what my brain could see.

I found that this really improved my situational awareness as well. Pilots had to have that ability, and I had to improve that skill if I were to find and see what I was looking for.

There are many people who develop some kind of 6th sense for their profession. They do the same thing so many times that they know what to look for. Cops are a good example. They will see something different or out of the ordinary and will recognize the anomaly immediately. I was not any different from them. I just realized what I had to learn. I then made an effort to do just that: learn.

The Bus:

In early June, there was a bus accident in Baie St Paul, east of Québec City, along the north shore. There was a steep hill and a right curve at the bottom. In fact, the north shore of the St Lawrence River was steep hills for a couple of hundred miles. The hills rose sharply to 2,000 and up to 3,500 feet right up from the river. At Baie St Paul, a river emptied into the St Lawrence River, which made for steep hills getting down from the high hills to the town.

A bus lost its brakes on the way down the hill, and the driver was unable to make the curve. The bus rolled a couple of hundred feet down the steep slope and killed almost 20 people. We were sent for aerial photos of the catastrophe and to make sure that all the bodies thrown from the smashed bus were found and picked up. An extra-large heavy-duty tow truck had to be brought from Québec City because there was nothing local that could pull a bus up the steep slope.

Infrared:

A couple of weeks later, there was a police chiefs convention in Québec City, and a lot of them wanted to see the city from the air.

That evening, a man showed up to demonstrate an infrared camera. It was great. From the helicopter, I could see which side of a car the exhaust pipe was placed and the number of occupants.

With a group of people outside, you could tell which ones were women. The infrared camera detected their breasts. They were slightly cooler than the rest of their bodies, and this was visible from 500 feet in the air.

There was also a police officer who had a steel pin in his leg, and that was easily detectable. Ringing metal detectors at airports today is common with people with artificial hips and knees.

Bank robbery:

There was another bank robbery on the south shore. I picked up the dog master and his German shepherd on the way there. The robbers were unlucky. A cop was on their tail fairly quickly, and they ditched the car and headed into the woods. We arrived, and after I dropped the dog and his master at the robber's car, I covered the area from the air. They couldn't move because of the helicopter, and they had to move because of the dog. They were soon caught. They only got $5,000.00 from the robbery, and that was not worth a couple of years in jail, plus the restrictions they had on them for the rest of their lives. One restriction is leaving Canada for travel to another country. The country that a person would try to go to would refuse their entry; they don't want criminals entering their country. Another limitation is employment. Many companies and all governments at any level would refuse to hire a convicted criminal.

There are very few robberies where the thieves were never caught. Cops are relentless in their search, and thieves are thieves because they are lazy and usually quite stupid. (There are exceptions) Prisons are not full of geniuses.

Chaudière Falls:

A few days later, I got a call to go to the Chaudière River Falls on the south shore of Québec City. This river starts just north of Maine, in the USA, and empties into the St Lawrence River. The Chaudière River falls are 115 feet (35 meters) high. This river often floods several towns in the Beauce area in the hills south of Québec City in the spring.

Two 14-year-old boys were reported missing in the area. From the rim of land near the falls, the two boys were spotted floating on the side of the river near the base of the falls. The two boys were pulled from the river and put on the stretcher, which I put inside. The Bell 206B helicopter was made to accommodate two stretchers, but I only carried one stretcher in the cargo compartment. I flew one at a time to the police post on the south shore. When the two boys had been loaded into the ambulance, they were taken to the morgue.

The problem was me. The smell of a dead body after it had been in muddy waters overnight was sickening. They were not in body bags to contain the smell but just wrapped in blankets.

Humans eat a lot of sugar, and the sickly-sweet smell of a drowning victim is exactly that, a sickening smell that is strong and cloying, a lot of that is caused by the sugar.

I was sick and nauseous for the rest of the day. I landed back at the Air Service and told the mechanics to wash and disinfect the interior of the helicopter as best as they could. I went home, stripped, and into a hot shower with lots of shampoo and soap to wash off the smell. My uniform was bagged for the cleaners, and the rest of my clothes went straight into the washing machine.

After that day, I either tied drowning victims on top of the floats or slung them out using my cargo hook. Fresh dead and not from the water was OK inside, but never again a drowning victim inside.

Gaspé lobster feast:

In the middle of August, we were called to go to the Gaspé. The dog master and his German Shepard were also told to join me in the search of a man who

74

had cracked, had a mental breakdown, and taken his car and drove into the nearby bush and could not be found. He was hiding. People had seen him run into the bush and he evaded all searchers. We arrived, and it took us several hours to locate him; he kept running and hiding. With the helicopter, I literally herded him towards the police. They caught him and the owner of the company who he worked for was so pleased with our efforts that he invited the group of searchers and myself to a lobster feast.

We were seated around a table filled with wine bottles and plates of lobster. We each had a small pail beside us for the shells. We all had two, some, 3 lobsters. The greatest lobster feast we all had in our lives. They were all large lobsters, and just a few miles away was a seafood place where the fishermen brought their catches, so lobsters were readily available.

Over the years, I often stopped there and bought shrimp, lobster, and crab. Free transport in the cargo compartment of the police helicopter.

My wife was astatic. Right off the boat was so much cheaper and fresher than any store in Québec City. We had an 18 cubic foot freezer and there was lots of room for seafood.

Marijuana:

Towards the end of the month, I flew a lot. Marijuana season was starting, and there were a lot of little pot plantations well hidden, except from the sky. The pot needs lots of sunshine. We managed to find several plantations, but fewer than we had hoped.

It was getting to early fall, and on the way there and back, there were large accumulations of Canada geese and Snow geese.

St Anne de Beaupre has a very large area along the shore of the St Lawrence River that is a Bird Sanctuary on one of the largest migration flyways in the eastern half of Canada. Millions of Canada geese and Snow geese stop on their journey southbound. They come from Greenland, most of northern Québec, and the Arctic.

The government has built walkways in the marches and set up hides for paying tourists to view and photograph up close some of the millions of birds that pass down the flyway.

I would fly past on my way east or home at a low level, and the noise of the helicopter would cause the birds to rise in huge clouds of geese, filling the sky below me.

The south shore had mud flats, and the geese stopped and fed there as well. On the south shore, there was a hunting season, and there were hundreds of camouflaged blinds and camouflaged boats filled with hunters during the hunting season.

Most of the hunters had 12-gauge shotguns, but there were a few 10-gauge shotguns. The 12-gauge shotgun has a range of about 45 yards but the 10 gauge has more stopping power and could be effective up to 80 yards depending on choke and pellet composition.

The geese would always take flight when I passed over. Whether that was good or bad for the hunters, I don't know, but after passing, the geese would usually circle back to land.

All the years I was there, the pattern was the same: millions of geese used the flyway every year, and the skies would fill with geese during their migration.

The excellent movie "Sully" happened because of Canadian geese, "Miracle on the Hudson."

Portneuf photo:

One day, I was flying around Portneuf, a small town about 20 miles west of the Québec airport. We stopped at the police post and picked up two cops. One asked if he could take some photos of a certain section of the highway where there was a long, slow curve with double lines painted on the road to prevent passing. No problem.

A couple of months later, I saw the cop again, and he told me that the reason for the photos was that he was Eastbound on the highway, and a car coming from the other way passed a truck. The truck was an 18-wheeler, and the other driver didn't see the cop when passing the truck on the curve, so the cop, coming from the other direction, was forced off the road.

Naturally, he was pissed off. He caught the driver and gave him a ticket for unsafe driving. Double the cost and the number of demerit points of a regular passing ticket, plus costs.

He took the cop to court and explained to the judge that he could see well enough to pass. Normally, the judges, being lenient, would let him plead to the lesser offence, but the cop showed the judge the aerial photo, and there just so happened to be an 18-wheeler on the curve in the photo.

He was judged guilty due to the photo, and the cop was happy when he told me about it. The word passed around fast, and aerial photos became a frequent request.

Mafia club:

In September, the cops from Montreal wanted to track the movement of some mafia members. They had information that a certain lodge north of Baie Comeau (260 miles or 420 km) from Montreal, which is NE and downriver from Québec City, was owned and used by mafia members.

Nothing exciting happened but it was a very nice place with a large area of exclusive fishing.

Ed and his helicopter from Montreal and I carried the investigators to the lodge. They found a guest book with several names and signatures of known mafia members. The cops were really happy and told us that the information was a big help in their investigation of the known members and their guests.

Tadoussac Fog:

Several days later, I got a call to go and find a lost fisherman east of Québec City.

The cops had found his pickup, but not him, about 30 miles west of Tadoussac, which is at the mouth of the Saguenay River.

I flew the 2 cops to the nearest lake, and after about ½ an hour, they carefully went up a bald rock. It was about 20 feet above the water. They could see his body about 15 feet below the rock in the clear waters of the lake. That meant divers had to be called. I had to refuel at the Coast Guard station in Tadoussac, then come back, and then fly to Québec City.

The next day we came back with a diver, and the weather was worse, a lot of fog over the St Lawrence River.

Again, I had to go to Tadoussac for fuel. However, this time, there was a breeze from the east and it packed fog on the west side of the Saguenay River. I could see one ferry at the dock loading cars and trucks and thought that they should both be doing the same thing at the same time. I could not see 100 feet, let alone the far shore where my fuel was located. Checking the map, I had almost 14 degrees of variation. The far side of the Saguenay was a compass heading of 30 to 35 degrees, which meant that I had to add 14 degrees to my calculated heading on the map. I flew about 20 feet above the waters of the Saguenay, knowing that 2 ferries crossed the river every ½ an hour. I did not have a schedule. I was scaring the piss out of myself, but I put 50 degrees on the compass and started crossing the river. Halfway across the fog cleared and I could see the other ferry was still at her dock.

I proceeded to the Coast Guard base and got a full tank of fuel.

I could not see a ferry from what I could see of the river. I took off and flew over the ferry dock. The ship was loading, so I had to assume the other ferry was doing the same. I used the reciprocal heading from my first crossing and hovered into the fog until I came to the shore on the west side of the river. Hugging the coast until the fog cleared up, I then went back to the lake and picked up my observer, the diver, and they strapped the body to one of the floats. (Remember, never inside again). I then flew back to Québec City, where the cops had an ambulance waiting for us at the Air Service.

The Fart

The next several weeks, I did about 5 or 6 searches, gave out lots of tickets, and flew over a demonstration by a bunch of farmers. There were also occasions for aerial photos. One such event I did was in Thedford Mines, where there was a large fire. On the way back, I was sent to Sorel to pick up a government minister. He was a short, fat, very fat slug of a person. I should have put him in the back seat because his weight put the helicopter right at the edge of the weight and balance limits. I knew burning fuel would lessen the problem, so I let him choose the front passenger seat. Big mistake.

On the way back to Québec City, the fat slug leaned forward, lifted the right cheek of his fat ass, and blew a big fart. It smelled almost as bad as the drowning victim. I quickly opened both windshield vents, the heater fan, and the sliding window of the pilot's door. To say the least, I was very happy to get to Québec City and get rid of the fat slug.

Winter was fast approaching. Québec City gets lots of snow, and with my wife pregnant, we didn't go out a lot. I bought a snowmobile. It was a fast beast, easily over 100 mph. (160 kph). I had scouted the trails with the helicopter and knew where most of them went.

Northeast of the city was a small restaurant that was only open in the winter. I had checked out the trails to get there in the helicopter and, one evening went there with the snowmobile. It was known for its great pizza. While there I met a guy called Francois, Frank is what I always called him. My French was getting better, and we were able to communicate fairly well. Frank couldn't speak a word of English. He knew the area fairly well, and I used to ride around with him. We got to be good friends and spent a lot of time together cruising the snowmobile trails. Frank's brother owned a pharmacy, and he used to work there and do a lot of deliveries, so he knew where everything was in a fairly large area around where he lived. He was also a volunteer game warden and knew the bush well for miles around. Both qualities were to prove very useful to me in the future.

Three Rivers, Ski:

I was sent to Three Rivers, then north, to find a 70-year-old man who had a heart attack while cross-country skiing. He had lain in the snow in the bush with his friends while someone called the police, then me to get him out of there. After he was loaded into the helicopter for transport to the hospital, the cop lost the man's pulse and started pounding on his chest to get him back to life. He was pronounced dead after his arrival in the hospital. According to the law, he could not have been pronounced dead in the helicopter or the machine, and I would have to be held in quarantine until after the coroner had made his report.

That could not be permitted for the helicopter, and myself would not be available for use until the report was finished.

That was also true with an ambulance. The patient is always pronounced dead in the hospital by a doctor.

Back in Québec City, it was time for the carnival parade. What made this year different was that I brought my wife along. She was pregnant, and so my daughter's first ride in a helicopter was before she was born.

One of the last events of the carnival was the canoe race across the ice-filled St Lawrence River.

The 8-man teams paddled and pushed across the ice flows to get to the other side, where they were handed a flag, which they tied to the boat and then worked their way back across the river to the finish line. From the helicopter, we had the best seats and witnessed all the mishaps that occurred to some of the racers.

I was kept quite busy in March and flew over 57 hours.

We also had a storm with freezing rain. My driveway was a sheet of ice, and I knew that when I pulled my car out from under the carport, it would slide down the sloped driveway. I needed to go out and buy several bags of sand and road salt. As I was looking out the window, I saw a salt truck starting to come up our street. I rushed to the fridge, grabbed a cold beer, then my coat, and slid down to the bottom of the driveway. As the salt truck approached, I got him to stop, then jumped up onto his running board below the driver's door. As he rolled down his window, I asked him if he would be so kind as to salt and sand my driveway. I handed him the cold beer. I happily watched him do a very good job sanding and salting my driveway.

Daughter:

Later in April, my daughter was born, and I was with my wife and the nurses during the delivery. It was amazing to assist in the birth of my daughter. The doctor was using forceps to help pull my daughter out while I held my wife's head in my hands. The doctor pulled my wife's head from my hands while pulling on my daughter's head with his forceps. I had to pass my arms under her arms and hold my hands together across her breasts. Then my daughter came out when the doctor pulled. It was an amazing experience to witness the birth of my child.

TT Straps:

A couple of weeks later, I was called to Mont Joli. It is far up the north shore of the Gaspé peninsula, where the mountains get quite high. Upon arrival, I was told that a group had gone into the mountains that were still snow-covered and one of the girls had fallen and broken her leg. Their location was not very accurate, and we had to get high and look for their camp. Crossing the hills there was a very strong wind with severe turbulence. My observer had to hang on to his seat with both hands with his seatbelt and shoulder straps pulled as tight as possible. It was the worst turbulence that I had ever encountered, but we found the camp. My observer helped to load the girl onto the stretcher and then into the helicopter. I flew the girl back, staying low and following a valley with little turbulence. I flew back to Mont Joli and landed at the hospital.

The girl was carried into the hospital, and I had to shut down and wait to get my stretcher back. Shortly afterwards, I heard a screaming anguish of pain when the doctor had to pull her foot to reset the bones of her leg. It was a while before I got my stretcher back. The doctor had left the girl on it to make the repair. When I got the stretcher, I went back to Rimouski to refuel and then on home to Québec City.

A few months later, I was called into the office of the operations manager. He showed me the TT straps from my helicopter.

TT straps are the key to the rotor head. Tension, torsion straps. They hold the rotor blade to the rotor hub. They are about 14 inches long with a large hole at each end to insert the bolts that hold the rotor blade at one end to the rotor hub at the other end. The strap is exactly that, thousands of very thin, very strong steel wires wrapped from bolt hole to bolt hole. They can rotate, which permits the rotor blades to rotate to add pitch to the blades for flight. The TT straps were a very formidable, very strong, very flexible piece of engineering hardware for the rotor blades. Many of the wires were broken on both of them. The mechanics who did the blade hub inspection were very surprised to see some wires broken. None of them had ever seen that before. The whole rotor head and the blades were magna- fluxed and given a very intense inspection.

Everything else was OK, but the TT straps had to be replaced. No wires, none, were permitted to be broken for the part to pass the quality assurance inspection.

Roberval fishing:

A couple of weeks later, I was called to Roberval on Lake St Jean. Two fishermen were missing. The cops had found their pickup, but no people. They were easy to find. A short portage to a small lake was where we found the 2 fishermen. They were both at the bottom of the lake beside the small boat. They had tied an anchor to the side of the boat in the middle of the oar lock. When they threw the weight over the side, the rope was too short. It capsized the boat. They had already caught several trout, and the trout were still alive, connected to the boat with a chain gill hook.

The fishermen died because they were stupid. The anchor was too heavy, the rope too short, they had tied the anchor to the oarlock in the middle of the small boat, and they were not wearing life jackets. Two more who looked at the sky through sightless eyes while tied to the top of my floats on the way to the morgue.

It is seldom that one thing will kill somebody, it is usually a combination of several things that are done wrong that kill. The 2 fishermen were an excellent example. Four stupid mistakes killed them. Remove any one of the mistakes, and they would not have died.

Over the next couple of months, I flew quite a bit. It was time for fishing and walks in the woods. I had to find several people, and there was always a drowning every month or so. I got to practice my search skills. I was getting better, didn't have to look too long nor did we leave anyone in the bush. We always kept at it until the people were found. Sometimes, they walked out themselves or other searchers found them. I was getting better. I learned more and more about how to search efficiently, and my 6th sense was improving, making our search times shorter.

Manic 3:

I was called to go to a robbery north of Baie Comeau, which is 260 miles northeast of Québec City along the north shore of the St Lawrence River. The robbery was from a Hydro Québec paymaster's office at the Manic 3 dam site.

Manic 3 is one of 5 dams on the Manicouagan River. Manicougan Lake was formed 214 million of years ago by a 3-mile-wide asteroid strike. The donut-shaped lake was created by the shock waves from the asteroid's impact and is the second-largest lake in Québec. Manic 5 Lake is about 1500 feet above sea level and Hydro Québec takes full advantage of this drop of elevation on its way to the St Lawrence River. Manic 5 dam alone generates 2,660 megawatts of electricity.

From Baie Comeau, I picked up a police observer and went to the Manic 3 reservoir Hydro Québec base camp.

It was from there that the workers were paid, in cash.

The three thieves stole $1,064,623.00. From the paymaster's office, they stole a truck and went to a nearby dock where they had a boat & motor stashed.

Their plan was to go to a cabin on the reservoir. There were many cabins on the lake and their plan was to wait until the furor died down and exit along with the many fishermen and tourists who frequented the huge area.

I refilled the fuel tank from Hydro Québec's fuel reservoir and was waiting to find out the plans for the search of the thieves when I received a phone call from the control tower operator at the Baie Comeau airport. The message was: "Your wife is in the hospital, and your baby is being looked after by the neighbour."

I immediately called the Air Service dispatcher and asked what they knew and what they were doing to help me and my family. The only information they had was what they had given to the Baie Comeau tower operator.

I gave them one hour to find out more information.

An hour later, I called the Air Service dispatcher and he told me they didn't have the time to find anything out because they had an emergency in progress where they had to send a jet up north to pick up someone who was sick. I growled back that I had an emergency, and I worked for the Air Service and that they should have taken the time to help a fellow employee.

I told them I was coming back. They balked at that, and I told them to go to hell, I was on my way and slammed down the phone. I fired up my helicopter, climbed to 6,000 feet (obeying NEODD SWEVEN) to get over the hills, and set course for Québec City. After landing, I filled out the aircraft logbook, tied the rotor blades down, got to my car, and headed for the hospital.

At the hospital, I went straight to the elevator. A security guard stepped in front of me to prevent me from getting on the elevator. He haughtily informed me that visiting hours were not for a few more hours.

I'm 6 feet tall, weighed 180 lbs, and was in good shape from playing a lot of racquetball and hockey.

I stepped closer to him and looked down at his 5 foot 8-inch height from only a couple of inches away and softly told him that he was too little to stop me.

I was wearing my uniform with 4 bars of a captain and looked imposing. I put two fingers lightly on his shoulder, stepped around him, and entered the elevator.

My wife looked wan and pale.

After I gently caressed her forehead, her eyes fluttered open, and she started crying. Big fat tears rolled from her eyes. "They cut me open," she lamented. My beautiful wife, so proud of her body, had to get cut open by the doctor to remove her burst appendix.

The nurse told me my wife would be there for a week at least with 2 drain holes in her abdomen to drain the poison from her burst appendix.

When my wife got to the hospital, the doctor did not know right away what the problem was. She later told the nurse that her abdomen was swollen like she was pregnant. She had given birth to our daughter only three months previously. That got the nurses' attention. She called the doctor, and he operated on her to remove the appendix less than two hours later.

When I left, she was still drugged up against the pain. I had to get home and pick up my daughter and thank our neighbour for her timely help in our hours of desperate need.

Emergency measures were called for.

I called my mother-in-law. I told her what happened and that I would be in Montreal to pick her up in three hours.

I packed up my daughter and all the needed accessories and headed for Montreal.

My mother-in-law was a rare jewel. She never questioned a decision of mine and was always available when we needed her. I told her several times over the years that I married her daughter so that I could have her as my mother-in-law.

The next day, when I went back to visit my wife, the same security guard was on duty and did not even blink when I walked past him to the elevator.

My wife was looking and feeling much better. She was happy to know that her mother would be with us for as long as she was needed.

Naturally, my daughter was spoiled rotten with all the attention from her grandmother, and me too, because I didn't have to cook or change diapers.

Back at the air service, I was told that I had to go back to Baie Comeau to finish the search. The next day when I got there, I took back the helicopter from my very pissed-off chief pilot.

The next day, I flew over 10 hours on the search. Every cabin, boat, and vehicle was checked out.

With the helicopter buzzing overhead, the robbers decided to leave earlier than planned and were caught during the night. The major flaw in their plan was that they did not have an airplane to get out of the area, the same day as the robbery.

I found out from the cops that the robbers had been caught and the money recovered. The cops kept me there for the rest of the day to complete the investigation and take a bunch of aerial photos.

Next day, I went back to Québec City.

The next week, my medical was due. All pilots need an annual medical from an aviation-certified medical doctor. Before 40 years old, the medical is required once a year, and after 40 years old, the requirement is a medical every six months with an electrocardiogram once a year. There are special rules for pilots, and the doctors who give the medicals must know them, and what drugs are or are not to be available for pilots. Giving blood is one example I mentioned earlier, and another is the mixing of certain medications.

I told the doctor what happened with my wife and the complete lack, on the part of the Air Service, to assist one of their own. I had left the job without permission.

I found out later that the doctor had talked to the director of operations and gave his opinion of the lack of any assistance from the Air Service.

I never again heard a word from my chief pilot or the director of operations about my actions when my wife was in the hospital. The cops didn't care; they still had the helicopter on duty, but with a different pilot.

The Trees

Contraband Diapers:

I had been flying a lot in the past few months, and one Sunday, I was called out for a local patrol. It was a rare weekend when there were no festivals somewhere. My wife told me before I left that she was almost out of diapers and baby powder. At that time, pharmacies were not allowed to be open on Sundays in Québec. There was only one in the Québec City area, and I didn't have time to go and get some diapers before I went flying, so I called Frank, who worked in a pharmacy owned by his brother. He told me he could get me what I wanted, and we agreed on a large open field not far from his place. He asked if I could take him and his wife for a little tour. I told him I would have to get permission from my police observer and if it was OK with him, sure.

As promised, Frank and his wife were there in his pick-up, waiting for us. We had nothing special for now, so the cop agreed to give them a little tour. Frank had a package of disposable diapers and a bottle of Baby Powder as promised. I paid him and he would leave them on the back seat when he left.

I gave them about a 15-minute tour of the Québec City area and over their house. They loved it.

When we got back to the field to drop them off near their truck, we saw there were three police cars in the field waiting near his pick-up. I landed near Frank's pick-up.

Two cops started approaching the helicopter, which was painted in Québec Provincial Police colours, plus it had a large police decal on each side of the fuselage. I gave a quick chirp from the siren to get their attention and motioned that I wanted only one of them to approach the machine. They obeyed, and Frank and his wife got out and went to meet the rest of the cops. They all knew Frank and his pick-up. He was a regular on the streets of his town, delivering stuff from the pharmacy, plus his job as a volunteer game warden. A town cop came to the door and talked to the uniformed Provincial Police officer in the left seat. All was OK, and we left to continue our patrol of the city and environs.

When I finished the day's flying and was back home, I called Frank and asked what the commotion was with the three town cop cars.

He could barely stop laughing while he told me a concerned citizen from about half a mile away had called the town cops. His story was a frantic telling of a helicopter meeting a pick-up truck and the truck driver had brought a large package of contraband to the helicopter.

This "concerned citizen" was a frequent caller to the local police to complain about anything that he didn't like. The very annoyed Orsainville police sergeant told the "concerned citizen" that the helicopter was from the Québec Provincial Police and the package (the contraband), was diapers for the pilot's newborn child. The sergeant had sent all his available forces that Sunday morning to intercept the "contraband," and when told that his cops had intercepted the Provincial Police helicopter with "contraband diapers," the sergeant was really pissed. He called and told the idiot never to call again and complain about anything unless his house was on fire. The idiot had called the police on several occasions before, and there were always false alarms, and with this incident, the sergeant had had enough of the person's BS.

Three in a boat:

The next weekend, we got a call to go just north of Québec City. A small empty boat was seen floating past a bridge with fishing gear hanging on the sides.

I found 3 people fairly quickly floating on the sides of the river. One was easily pulled from the river and loaded into an ambulance.

The other two were on the other side of the river which was a jumble of weeds and floating debris. I couldn't get closer than 10 feet. The cop checked, and the emergency squad divers would take several hours to get there, so I got on the police radio and made a conference call to Frank to ask him to get what we would need to get the bodies. I knew he could get a small rubber raft that could get to where the bodies were stuck.

We waited awhile, and when Frank showed up, we explained what we needed.

He had a rubber raft and lots of rope.

He had to cross the river and wrap the rope around the ankles of each victim one at a time, then get to the far shore. He then had to get far enough away from the raft so my rotor downwash would not blow the boat around. He then had to clip a loop from the rope to the cargo hook of the helicopter while I hovered over him.

I then slung the body back to the other shore and the waiting ambulance. I had to do this twice; then Frank was able to come back to collect his ropes and a swift goodbye from us. It had been a gruesome job with a lot of spectators.

That evening, I went to his place, and we discussed the various things that I could ask him to do. He was enthusiastic and enjoyed helping us in spite of the morbidity of the last job.

Cigarettes and a boater:

I went to Sherbrooke to try to find two trucks full of cigarettes that had been stolen. I guess the thieves were smart and hid the truck in a barn or garage. They were not seen and I didn't find them.

From there, I was sent to Schefferville to find a missing boater.

Schefferville is over 4 hours flying north of Seven Islands on the lower north shore of the St Lawrence River.

Seven Islands is 11 hours driving NE of Montreal. Schefferville is almost 7 hours flying NE of Québec City through Seven Islands.

The one thing about a helicopter is that your route of flight is always governed by your need for fuel. One of the problems with flying up the Moise River north of Sept Iles (Seven Islands) is that the route of flight was up the river following the railroad tracks. It was the best and safest route to Schefferville, except for one place that was really dangerous. About 50 miles north of the St Lawrence River, if following the river, is a radio tower. There are lots of them on the way north. This one was so close to the tracks that one of the guy wires holding the tower vertical was strung across the tracks. It was a small steel wire and nearly invisible. A low- flying helicopter in foggy weather could easily hit that guy wire. Deadly dangerous. All pilots had to be told about it and have it marked on their maps. The wire should have had red ball markers attached to it to make it visible, but did not.

They told me what route the boater was supposed to be, and I found him on the shore of a long lake. He was lying on the beach, and his skin was burned black. He was disgusting to look at. No way were we going to touch him and try to load him on a float. We didn't have a body bag. No way was I going to get near the body either; the smell would have made me sick for the rest of the day. We went back to Schefferville and sent a crew of Indians to cover him with branches or a tarp to keep the animals and birds off him.

The next day, they went back in with a body bag and several masks that had to be flown in on the morning flight from Seven Islands.

On the way back to Québec City, I had to spend several hours east of Tadoussac looking for some suspected marijuana fields.

Young and old:

The next couple of days, it was searching for lost people. The first was a 7-year-old boy in the bush. It took several hours to find him, but he was OK and was happy to get a helicopter ride.

The next day, around the fields and swamps southwest of Québec City, an 80-year-old man had gone for a walk after breakfast, and when he didn't return by early afternoon, his family called the police. This guy was easy to find. He had wandered into a large area with lots of gravel and dirt roads in a huge swampy area. He was really excited when we stopped to pick him up and take him home. It was his 80th birthday and the first time in his life that he had ever flown in anything.

Pot Hunt:

We went to Riviere du Loup (Wolf River) east of Québec City on the south shore.

This one was worth it. I found a large area of half harvested pot plants, and the cops got a search warrant for the house that had the entrance road to the field of pot plants right behind it. The house was full of marijuana, and you could smell it 50 feet from the house. There were 3 rooms filled with heaters, fans, and drying pot plants. The living room had several dozen large one-pound Ziploc bags full of pot ready to sell and a dozen pairs of scissors. The dried and ready-for-market pot from the field and the house would have totalled several hundred pounds of very valuable pot. There were also a few small bags of black hash, worth far more than their weight in gold, scraped from the scissors.

It took a while to learn how to spot pot fields. Once I did, they were much easier to find. In fact, I found more pot fields flying for Bell Helicopter than I ever did for the cops.

Swimming and car races:

There are 2 notable races in the Lake St Jean area. One at the end of July and the other at the end of August. The swimming race at the end of July is in Roberval and the Labour Day weekend had a large 3-part stock car race in St Felicien a few miles to the northwest near the St Felicien Zoo. The St Felicien Zoo is huge and world-renowned with many thousands of visitors every year. I later visited the zoo with my family, and the place had lots of people and the parking was full.

The swimming race is 20 miles or 32 kilometers across Lac St Jean. There are 25 of the world's top distance swimmers that compete. The purse was then $40,000. It has now risen to over $55,000 shared over the top few swimmers. It usually takes about 7 hours to cross from the north shore of the lake to Roberval on the south shore. The start was 8:30 AM, and a boat followed each swimmer. Hundreds of other boats were also buzzing around the swimmers. At that time, the cops did not have a police boat to monitor the drunken boat drivers. A lot of beer was swilled on the boats also. Every hotel, motel, rental room, campground, and cottage was full of spectators for the race.

Labatt Brewery had the franchise, and they brought in a dozen train box cars full of beer for the long weekend event.

Scavengers made a lot of money collected from the empty beer bottles they turned in at 0.10 cents a bottle. The hundreds of lawns and all the garbage pails placed around the town were picked clean of all bottles by early morning.

When the trains left, they were again full, with empty beer bottles.

The cops had one whole motel booked for themselves, so I had no worry about having a place to sleep, and I was chauffeured everywhere I went in a police car.

For the swim race, everybody was packed into and around the town, and almost everything was within walking distance. Not so for the car races. All the motels were around Roberval with only very few small motels near the race track in St Felicien.

The first race was the stock pick-up trucks. The second was the stock car race. The third was a demolition derby. This one was last, of course, because of all the time it took to clear the oval track of all the wrecked vehicles after the race.

For the swimming race, I crossed and recrossed the lake following the swimmers. We also allowed a reporter to fly with us and use a police radio frequency so his reporting could be monitored by the police. This let the Roberval radio station able to listen to and report the leaders of the advancing swimmers a lot more efficiently than what was done on one of the following boaters.

The car races were hugely different. Nobody could walk there or use a boat to be a spectator. Everyone had to drive there, and all the roads were two lanes with many curves and intersections. There were a lot of areas where there were double lines to prevent the legal passing of another vehicle.

Many drivers wanted to race to the races and, after seeing them, had the racing fire in their veins and wanted to race back to where they were staying.

We caught over 70 drivers with the helicopter for breaking the passing zone laws, and some were caught more than once by police cars stationed along the way.

There were no head-on collisions that weekend between Roberval and St Felicien and no dead people.

The local media made a big thing of it because it was the first time in years that there had been no major accidents or deaths to or from the car races.

St. Tite Western Festival:

If ever there was a cowboy or cowgirl in Québec, this is where you would find them in September.

North of (Three Rivers) Trois Rivières along the St Lawrence River is the small town of St Tite. It is a mini-Calgary stampede in the middle of nowhere.

It got started because of Boulet Boots, a maker of cowboy boots and dress boots.

I was again there for traffic control for the weekend. I slept at a motel in the relative quiet of Trois Rivières and passed the days at St Tite. The town was pure Western, minus the pistols.

I bought my first pair of cowboy boots in Hong Kong and later in Fort Worth, Texas. Here was perfect. Excellent quality cowboy boots at a manufacturer's price.

The police offered the owner a ride in the police helicopter with his family. I was happy to do it. We took them for a ride all around the area.

Each year, I would come back to the festival with the helicopter, and the owner of Boulet Boots sold me and all the other cops quality cowboy boots very cheaply. Every year, I would buy a couple of pairs of boots, dress boots, cowboy boots, and walking boots. Anything I wanted for a very low-cost pair each year.

To this day I still have a couple of pairs of Boulet cowboy boots that I have not worn out.

Boots aside there was almost everything you could find at the Calgary Stampede or a Fort Worth rodeo, only on a smaller scale. The cowboys roped calves, got bucked off bulls, and there was nary a Texan among them; they all spoke French.

On 2 nights during the long weekend I was there, the cops had me leave the helicopter in the gated, secure area at the minister of transport compound, take me to my motel to change out of my uniform, and then bring me back to Ste Tite for dinner and the evening's festivities. I'd have dinner with several of them, and a couple of beers, then they would drive me back to Three Rivers to sleep. The second evening I met a cop who had spent 6 years in Montreal on a special squad that would follow criminals around. They tried to do this discreetly, but they used several cars during a tail, and there were no traffic laws for them. All undercover, and all with amazing driving skills. Most would wreck between 5 and 10 cars a year, but they wouldn't lose their suspect.

This cop offered to drive me back to Three Rivers, the fast way. He didn't use the main highway. He took the back roads, the gravel roads, seldom below 100 mph. Every corner or curve was a 4-wheel drift. His driving skills were simply amazing. The engine was smoking when we arrived, but we did arrive safe and sound, just very damn quickly.

Cap Trinity:

We were called to Cap Trinity to pick up a man who had died of a heart attack after he had climbed the statue at Cap Trinity. It was a very long walk up a 1300-foot hill to get the body and then bring him back down the thousands of steps to get him to an ambulance.

The rescuers were to bring him straight down the very steep hill to the Saguenay River at the base to load him into the helicopter. No steps, but much shorter to go through the bush down the hill to the river.

Cap Trinity has an interesting history. In the 1880's a man fell through the ice walking on his way to Chicoutimi. In the freezing waters, he vowed that if he lived, he would build a statue at the top of the hill at the point of Cap Trinity. He did both. The statue is Notre-Dame-du-Saguenay. It weighs 9 tons, is 30 feet or 9 meters tall, made of pine sheathed in thin lead sheets.

It stands to this day and is a small provincial park.

Aircraft crash:

The Canadian Armed Forces search and rescue squadron from Trenton, Ontario, found the aircraft using its ELT (Emergency Locator Transmitter) by law required in all Canadian registered aircraft. The Québec Police were tasked with removing the four bodies. Deep in the woods, midway down the Gaspé peninsula, a group of men went in to cut enough trees for my helicopter to land.

Laying each body on the helicopter stretcher, I removed the bodies one at a time out to an ambulance, appropriately parked in the large parking lot of the nearest town's church.

The fourth one was really gruesome. His head was cut off and placed between his knees on the stretcher. The body was found upside down and suffered from exsanguination. A fancy way of saying that he had no blood left in the body. The average man has about one and a half gallons or six quarts of blood in his body.

He was a lot more than just a quart low; he probably didn't have a quart of blood left in his body. (Imperial quart = 1.13 liters. US quart = .946 liters or 1 liter = 1.06 US)

(An ELT is a small portable signal radio that transmits an emergency locator beacon on 121.5 [VHF] and 243.0 [UHF] emergency radio frequencies and to overhead satellites.

(VHF=Very High-Frequency commercial aircraft frequency band, and UHF=Ultra High Frequency, military aircraft band radios.)

Bear

The Orsainville zoo was across the highway from the Orsainville prison, where they each kept their various types of animals in cages.

A small bear escaped from the zoo and was scaring the nearby citizens. We had to catch him alive.

I had to go back to my base and get a large cargo net for the cops to catch him. I herded him to the net holding cops, and they managed to catch and tie him up.

He was a full-grown bear but a runt, about half the size of a normal adult bear, and he was just as strong.

They finally got him tied up, and he was put in the trunk of a police car and taken back to the zoo. He was dead when they got him there. He died of fright or a heart attack in the ten-minute ride in the trunk of the police car back to the zoo.

We had several Orsainville cops and also several Provincial cops to herd the bear and protect the people stupid enough to come out and watch. The problem was the zoo people couldn't get someone there with a tranquilizer rifle there quickly enough. That cost the bear his life.

Porche:

We spent a lot of time giving speeding tickets. This method was new in Québec and was very effective using the painted triangles 500 meters apart and an aircraft or a helicopter.

On one occasion, we were working just south of Québec City on the autoroute, and a cop car from about 40 miles west reported a Porsche moving very fast eastbound, right toward us.

I climbed up a couple of thousand feet and spotted the speeding car from several miles away. My observer advised the cops on the highway, and they simply stopped all the traffic eastbound. We timed the Porsche at well over 100 miles per hour.

The Porsche slowed when the driver saw the traffic jam and he was then herded to the shoulder of the highway. The cops reopened the highway to let the stopped cars go on their way and spent a long time checking the car and the driver of the Porsche. He left a long time later after being thoroughly checked and with some very large numbers on his speeding ticket.

For a lot of these rich and 'to hell with the cops' type of people, what really slowed them down was the demerit points. If they lost their license on points, they had to go back to the same routine as a 16-year-old. Driving classes, driving lessons, multiple restrictions, and very high insurance rates.

A few years later, the Cadillac Man is a case in point.

A couple of days later, I was working east of the city, and I heard a cop on the radio calling Québec for license information from the park 50 miles north of the city,. Several minutes later, he called and told other cops to be on the lookout for a certain car heading for Montreal, the driver was in a great hurry.

About 30 minutes later another cop called with the demand for the same car from the park. Later, the cop advised Québec metro to be on the lookout for the same car. Metro missed him, but he was later caught on the south shore west of the twin bridges, still speeding towards Montreal. After three tickets and 7 to 9 demerit points, I hope his haste was worth it.

Some people don't realize that cops are people too and know how to think, plus they have a radio to forward information.

Hangar talk:

One afternoon, the Emergency Squad Captain decided that all his emergency squad people should be familiar with where the helicopter was kept. I met with 2 shifts of the squad at the hangar, and they started asking me questions about the machine and its operation. I showed them around and in the same hangar were the executive and air ambulance aircraft. The Canadair CL215 water bombers were also there, and this was their home base. The Air Service had about 15 of the CL215s for fighting forest fires, and of course, all the pilots for all the aircraft were also based in Québec City.

One of the cops asked me why I was always looking for something to check out, give tickets, or search around sand pits and quarries. Why didn't I just fly around and take it easy?

I told him that if I spent seven days just sightseeing we would be wasting a lot of time and money for nothing. He said he preferred to just fly around and patrol the city. I told him if that's what you want to do, do not come flying with me. Go fly with the other pilot. He was taken aback by that and didn't say another word.

The other cops got the message loud and clear.

After that, the police observers who wanted to do something were the ones who showed up to fly with me. I never flew with that cop again.

More trees

All over the world, there are thousands of airports. The United States has the most airports of any country in the world. Every airport has a four-letter identifier.

In the continental USA all start with the letter K. KLAX or LAX is Los Angeles International Airport. New York is KJFK or JFK, John Fitzgerald Kennedy Airport in New York City. All US airports start with K, except Alaska and Hawaii.

Bombay, India's ADF navigation beacon is BBB (OBBB), even though the city's name has now been changed to Mumbai. The are almost two dozen different starting letters for different areas of the world. Russia is U, China is Z, Australia is Y, etc.

Normally, only the last three letters or numbers are used. London, England, has ELHR for Heathrow and ELGW for ELGW for Gatwick. The E or any first letters are not normally used on flight plans.

The ones that most concerned me were the Canadian airports, which all start with C. Montreal is CYUL, or just YUL. Québec City is YQB, Winnipeg is YWG, and Toronto International is YYZ. The codes do not always seem logical to go with the name of the airport. YZ comes from the old railway code for Malton railway station and the original Malton Airport, which is where the airport is located.

The one that will concern me next is CYKL, Schefferville, in far northern Québec.

Growing

Black boot:

I was called to go to YKL, Schefferville, to look for a man who went cross-country skiing and never returned. He had departed in the morning, and in the early afternoon, a snowstorm arrived and turned the nearby lakes into frozen fields and the blowing snow created whiteout conditions with extremely reduced visibility.

I left YQB, Québec City, flew to YBC, Baie Comeau, for fuel, then on to YZV, Sept Iles (Seven Islands) again for fuel, and then north to YKL, Schefferville.

I picked up a police observer and we started searching over the several small lakes in the area. The lakes were dotted with little islands, and on the lee side of some of these islands, we saw ski tracks. The problem was that as soon as he skied back into the wind, there were no more tracks. I was doing this for quite a while when I saw the tip of what looked like a small fist-sized black rock. The rocks were all grey around there, so I went in for a closer look. I hovered myself into a whiteout condition, keeping the shiny black spot as my reference point. I kept blowing the snow from the black spot for several moments until the bootlaces slowly became uncovered. I rose vertically to get out of the whiteout condition, and then we marked the spot on the map, called on the radio to get a pick-up crew together on a couple of snowmobiles. Asking them to bring one, including a sled, to carry back the frozen victim.

It was a long way to fly to find the toe of a boot. I had been trying to train myself to find the unusual and this time it worked.

The next day, I flew the long flight back to Québec City.

Moving day:

This was a busy year. I flew a lot, but mostly routine. Searches, Québec carnival, giving tickets, sled dog races, canoe races over icy rivers, which were fun to watch, and lots of aerial photos.

In the late winter, my wife and I decided to buy a house and get out of the apartment.

Moving day was a snowstorm and not a fun day to move.

The last thing I moved was my snowmobile. The next day, with the snowmobile still in the trailer, I took it to a dealer and traded it in for a new snowblower.

My driveway was 90 degrees to the snowbound winds, so the smallest snowfall would quickly fill my driveway with snow.

We soon met our neighbours. Mike, across the street, was using a shovel on his driveway snow. I offered to help him with my new snowblower. Soon, the wives met, and then we met a few of their friends, Jack and Pete.

Pete and I did a lot together and are best friends to this day.

Police dog:

The Québec police emergency squad in Montreal started using German Shepherds to assist in search, rescue, theft, and whatever else they did with police dogs.

The first time I worked with a dog was in Shawinigan, north of Trois Rivières, a city between Montreal and Québec City.

Three young boys had wandered into the woods from a campground and had not returned in the evening.

The dog was called from Montreal, and I came in from Québec City. I decided to search in the same direction as the dog's nose. Scanning far ahead of the dog named Grimm, the boys heard the helicopter and found a small lake where I spotted them.

We called the dog master and told him that we had the boys and met him with his dog back at the campground.

One thing I found is that not only does the dog master have to train the dog, but the dog master has to train himself to read his dog's actions.

We had a lot of searches in the eastern half of Québec, and we later had a dog master assigned to the Québec emergency squad.

Quite often, when I went to search for someone somewhere, the police on the scene would have some local man show up wanting to help with the search in the area, which he said he knew the area extremely well.

It took me a while to realize this, but they would know the area from the ground, horizontally. When I would get them in the air, they had a hard time orienting themselves. I'd have to go to a lake, river, or road for them to orient themselves. I could take someone up, go around a hill, make one or two turns to read the land and they would be lost. They were not used to looking at what they knew so well horizontally to the newer view, vertically. So, after a while, I realized that they were more of an inconvenience than any help. Plus, the added inconvenience of them getting airsick. They would barf out the small passenger door window and make a mess or use an airsick bag mounted prominently on the center column facing them with the resulting offensive stink.

Another thing I learned was that, when people who got lost in the hills went up when they did not find their way out themselves. When you climb a hill in the bush, all you get is more trees and when they get to the top, they can't see to orient themselves because they can't see through the trees below them. If they don't have a map and a compass, they need to go down. Down, you will find the lakes, rivers, roads, cottages, in short, other people who could help you call home or the cops.

I used to tell this to my wife.

We owned a cottage north of Montreal, and one time, she went there with my parents and my brother. They went up the road to another piece of land I owned and went for a walk in the woods. They promptly got lost, and my dad wanted to go up to see better. My wife told them what I had told her, and she promptly turned and went downhill. The others protested but followed her. They finally came out of the bush near sunset and saw that they were only a few hundred yards from our cottage.

For all, a cold beer tasted good.

Kayak rock:

The Chaudière River Falls is only a short distance south of the St Lawrence River. This is where the two previously mentioned 14-year-old boys I lifted from the river were found.

An intrepid adventurer carried his kayak up to the base of the falls with the intention of floating down the rapids on the short ride to the St Lawrence River. He badly misjudged the turbulent violence of the water coming down the Chaudière Falls at least he was wearing a wet suit. An observer from the rim of the falls spotted the man on the rock and called the cops. I never saw the kayak, it was probably halfway to Orleans Island by then.

I picked up an observer just south of the Québec twin bridges in St. Nicolas and went the few miles to the falls.

The person was in a bad position for me. On the west side of the river were sheer rock walls about 100 feet from the rock he was sitting on in the rushing rapids. He was more than 300 feet from the east side, where there was a small grassy slope. From both places, it was impossible to throw a rope that far. The round water-worn rock with him draped over the top was about 8 feet in diameter, and there was a drop of about 3 feet right where the rock was located in the rapids. The wind was from the mouth of the river, opposite the water current. I had fixed floats on the helicopter but could not land on the fast-flowing water with my tail into the flow of water. I couldn't get low enough to let him climb onto the float because that would put my tail rotor into the flowing waters. Everything was wrong for me to get to him, water flow, wind, and no places to set down in the violent rushing waters. He had to grab onto the crosspiece of the float bags.

I had to hover into the wind facing downstream. As I hovered closer to him, I took my hands off the controls for a second or two and motioned with my hands for him to grab the crosspiece. I hovered right up to him and put the crosspiece inches from his face. He grabbed onto the crosspiece with his arms, and I slowly lifted him up off the rock. I could not see him because he was directly under my seat, hanging on. I hovered to the east side of the river and lowered him onto the grass. I felt his weight leave the machine and he then moved forward where I could see him. He was OK, scared shitless, but OK.

A cop was there to receive him and took him to the police post for personal information. I dropped off the observer and flew back to base.

I was later told that the guy was more scared of hanging onto the helicopter than he was hanging onto the rock. I don't blame him; the rock was solid and his life was literally in my hands and his ability to hang onto the round pipe of the float crosspiece.

212 and 412:

There was another plane crash on the lower north shore about 100 miles east of Québec City. The airplane was an Apache, and the crash killed five people. The same recovery routine as the first aircraft crash.

Search and rescue from Trenton, Ontario, sent a Hercules aircraft, as usual, to find the missing aircraft. The Québec police were alerted, and I went to assist in the recovery of the bodies. The military sent a twin Huey to assist. They couldn't do too much at first because they had no air-to-ground communication. They had to go through me to talk to anyone on the ground, plus the pilots were from out west and couldn't speak French.

We coordinated our efforts, and they helped a lot with their larger machine.

There is an interesting fact about the twin Huey.

After the Vietnam War, the Canadian military went to Bell Helicopter to buy some Huey's, but they wanted them to have two engines because of the off-shore and far north use of the helicopters.

Bell told them to pay for the development of the twin Huey, and Bell would pay a royalty to the Canadian government for all that they sold. They would have Pratt and Whitney twin-pack engines built in Montreal. This was to later include the 412-model helicopter which is still in production at Mirabel in 2024.

Bell has sold a few thousand 212 and 412 helicopters over the years to both military and civilian buyers.

Bell is still building the 412 with the upgraded twin-pack engines from Pratt and Whitney to this day, and paying the royalty to the Canadian government.

Canada bought 100 twin Hueys back then. In 1984, Bell moved civilian manufacturing to Canada, now in Mirabel, north of Montreal, Québec. Canada bought 100 412 helicopters in the 1990's, and called them the CFUTTH. Canadian Forces Utility Tactical Helicopter. Right now, they have 82 of the CH-146 Griffon helicopters left in service. The first one they lost was along the Labrador coast. The pilots got lost and landed on the ice close to shore, where they saw some buildings. Hoping to ask someone where they were and hope they had radio communication available, the pilots shut down the engines, and the helicopter promptly broke through the ice and sank in about 100 feet of water. The 2 pilots and 2 crew members managed to get out of the water and met the people in the small village to get warm and out of the cold.

A few months later, the military contracted a Newfoundland helicopter company to come with a diver and a S-61 Sikorsky helicopter to lift the CFUTTH 412 to the surface and put it on a barge. The chief pilot decided he wanted the glory, and instead of getting his most experienced long-line pilot to do the job, he did it himself. The chief pilot managed to get the 412 out of the water and was carrying it to the barge when he got going too fast. The 412 started oscillating under him and with his lack of experience doing long lines, he lost control of the 412. The chief pilot punched off the 412 and left it to drop in water over 1,000 feet deep. Unrecoverable.

Bell Helicopter hired a S-61 long-line pilot from that company (not the chief pilot) to become a test pilot at Bell. He told me the story of that mishap.

Montreal is now one of the only cities in the world that produces the whole list, top to bottom, in aircraft manufacturing. Simulators, (CAE), engines, (Pratt and Whitney plus Rolls Royce), Bell Helicopter, Bombardier, and now Airbus with the A-220 series aircraft. There are also hundreds of small companies that manufacture the parts for all these builders of simulators, engines, helicopters, and aircraft. There are also schools that train aircraft mechanics to work in these companies and universities that produce engineers for the aviation industry.

Every single part has to be certified by the US FAA and the Canadian MOT.

(United States Federal Aviation Administration and the Canadian Ministry of Transport)

Québec has long wanted to separate from Canada. Ref: Jacques Parizeau, 1995. If that ever happened, every single company that has FAA and MOT certification would have to move out of Québec or the manufacture of the parts would have to move out of Québec. There would be a massive loss of jobs in Québec. Just Airbus employs over 3,000 people plus all their parts manufacturers. Montreal's economy would be destroyed unless Montreal separated from Québec.

La Tuque strike:

I woke up to a clear, hot, sunny Saturday morning. The phone rang. The Three Rivers district needed the helicopter in La Tuque, a small town halfway from Three Rivers on the St Lawrence River to Roberval which was on Lake Saint-Jean.

The local lumber mill employees had decided to strike.

They set up a blockade on Highway #155 at the best choke point. It just happened to be across the road from the lumber mill. The hills came to a sheer drop only a couple of hundred yards from the Maurice River gorge dam.

The perfect place for a blockade. Only the main road, a railroad, a sandy parking lot for the mill, and the mill on the other side of the road had room to exist.

I arrived at the small La Tuque airport and refueled. An observer and I circled the area, and the strikers were massed along the highway with no other route around them. The cops sent in a dozen reinforcements and the situation was getting violent.

The cops needed a distraction. The weather was great. Hot, cloudless with light winds; perfect for me. I came to a high hover directly over all the strikers' vehicles in the parking lot. This was Québec in the early 1980s, very few vehicles had air conditioning. The few convertibles all had their tops down. Most of the cars and pick-up trucks had their windows open. It was a hot summer day, and the strikers were within sight of, and close to their vehicles, and controlled the access entrance to the sandy parking lot, so they were not worried about their vehicles.

From a high hover, the downflow of my rotor wash was easily strong enough from a 3,000-pound helicopter to pick up much of the debris, sand, and wood chips littering the ground and blow them into all the open windows of the vehicles and with plenty left over to thoroughly coat the insides of the convertibles with sand and debris.

The infuriated strikers left their places in the line of protesters on the highway to try to protect their vehicles. A couple of them picked up rocks or sticks to try to throw them at the helicopter. The rotor downwash was easily strong enough, and I was high enough to defeat any attempt. The only thing they could do was close the windows of their vehicles and be left with a big cleanup job in the interior of their vehicle. This distraction gave the cops the superior strength to break through the blockade and hold off the remaining strikers.

This did not end the strike, but the cops could now control the passage of the traffic, and the strikers were backed to the entrance gate of the lumber mill.

Mission accomplished.

21st Olympics:

Queen Elizabeth II opened the 21st Olympics in Montreal, Québec.

It was a rare honour for Canada to have Queen Elizabeth II in the country and also for Canada to host the Summer Olympics in Montreal.

I was with the police and was issued a security pass for the Olympics. It was an opportunity of a lifetime, and I took advantage of it.

The helicopter was kept busy in Québec City, and on my week off, I went to Montreal and stayed at my parent's place on the south shore. I took the new Montreal Metro downtown to the Montreal Forum. With my security pass, I could enter any Olympic venue I so desired. I went to the Montreal Forum. I did not have a seat, but I could stand anywhere that was not in someone's way.

I witnessed Nadia Comaneci score her first 10. It was an odd few minutes when she finished the balanced beam. It took several minutes for the score to be lit onto the scoreboard. Everyone was stunned. Her score was 1.0.

She was given the first perfect 10 in Olympic history.

The announcer came on in both languages and explained that the score was a perfect 10. The crowd roared in appreciation. It took several minutes for the cheering to abate, and the announcer explained that the scoreboard was only designed to go to 9.99 points.

Over the next few days, I witnessed 2 more perfect 10s from Nadia.

That little 14-year-old Romanian became the instant heroine of the Montreal Olympic games, and I was there to witness several of her medals. She was immediately nicknamed "the goddess from Montreal" by the media.

It was chilling to witness such a historic event, perfect 10s at the Olympics.

Never before and never again did or will that happen, the Olympic judges changed the scoring to divide the performance into two categories: artistic value and performance value.

The big "Owe":

The Montreal Olympics had three problems.

-The 1972 Munich massacre of the Jewish athletes.

-Massive corruption and mismanagement.

-African boycott.

11 Israeli athletes were murdered by the Palestine Black September group at the end of the 1972 Munich, Germany Olympics.

This resulted in a massive increase in security personnel, vastly increasing the cost of the Olympic games.

Example: I was the helicopter pilot for the Québec Provincial Police in Québec City, and I got a security pass. Those like me were not crazy, we used the passes to enter any Olympic venue we so desired. One good thing was it produced more visible security people everywhere but at an inflated security cost for the city.

Massive corruption. A huge example was cement. The Olympic stadium is massive and took millions of tons of cement. Cement trucks used to enter Olympic sites, get logged in, and the city would pay for the cement, but the mafia controlled all the construction companies. Many of the cement trucks would, after logging in, just turn around and drive back out of the site, go somewhere else, and dump their load in some other construction site where they would get paid a second time for their load of cement.

For years, there was a mafia boss in Montreal known as "Mr. Three Percent." The mafia got 3% of every construction project in the Montreal area. Violent consequences were the result of non-compliance.

Mayor Jean Drapeau of Montreal once stated that the Olympics would not cause a debt any more than a man could have a baby.

The total debt load for Montreal after the games was $1.5 Billion. It took 30 years for that debt load to get paid off from the increased taxes on cigarettes in the Province of Québec. Hence, the "Big O," the Olympic stadium, became known as the "Big Owe."

After that came out, every cartoon and caricature of Mayor Jean Drapeau showed him to be heavily pregnant.

28 African nations boycotted the Montreal Olympics because the IOC, (International Olympic Committee) permitted New Zealand to participate in the games. The reason for the boycott: The New Zealand "All Blacks" soccer team toured the then apartheid-era South Africa.

Lost:

I went up to Chibougamau for a search. Found two with a broken boat. Sent in a team to fix their problem. The next day, I flew back and was called to go to Chicoutimi for a search. It was a man alone who had camped on a small lake and got lost walking around a tall hill. He came to a small pond held back on the edge of a steep descent by a beaver dam. He then did exactly the wrong thing. He climbed the mountain instead of following the water, which would have taken him right back to his campground. He heard the helicopter and made a small fire halfway up the hill. I dropped off the dog master and his dog. They met the man coming down and we joined up, and we took him back to his camp. A cop car was there to talk to the man. I was to remember this hill a few years later with a sergeant.

Guy Sampson:

I was on my way to Baie Comeau and heard that Corporal Guy Sampson was shot.

I was called back and was the second police vehicle to reach the scene.

Cpl Sampson and another officer were looking for a man who had shot and killed a woman a few miles north. The man owned a little cabin NW of the Québec Airport.

The two police officers had gone to the cabin and when Cpl Sampson knocked on the door the suspect fired his .303 rifle bullet through the door. The bullet caught Cpl Sampson in the neck. He died instantly from the hydrostatic shock to the brain, but we didn't know that. A .303 bullet leaves the muzzle of a rifle with 2,554 ft/lbs of energy at 2,770 ft/sec with a 150-grain bullet. The effective range of the bullet is 7,000 yards, with an aimed killing range of 1,200 yards. Guy Sampson was two feet from the muzzle of the rifle when hit by the supersonic piece of lead from the .303. The hydrostatic shock exploded his brain. It's like hitting a fresh egg with a hard-swung golf club. Guy Sampson, Guy like we or he in French. Guy like high or why in English.

Corporal Guy (gee) Sampson was dead was dead so fast it took several seconds for his body to find out.

Two more cars arrived. Guy was the boss in charge of the district that Sunday morning. When I got there with the dog master, they wanted me to fly over the cottage and try to see if I could see Cpl Sampson's condition. The suspect had already killed a woman and Guy Sampson, and I knew he would not hesitate to shoot at my helicopter.

The cabin had its only door facing north and its only window on the west side. I had dropped off the dog master he wanted to sneak in from the west and see if he could see Cpl Sampson. The dog master was slowly creeping through the bush when a bullet missed his right ear by inches, and the shock of the passing bullet knocked him on his ass. I was supposed to distract the murderer, so at the same time, I was coming in at tree-top height and at maximum speed to see if I could catch a glimpse of Cpl Sampson by the cabin door. As I passed over the cabin, I rolled hard left and kicked the left pedal to turn as fast as I could, thereby letting torque help me by rolling in the direction of rotation of the rotor blades. The .303 bullet passed just under the chin bubble of my helicopter. It may have missed by only a few inches. I knew from my combat experience the sound of a very closely passing bullet. That is one sound that no person can ever forget. I even have a tape recording of a time when we were being shot at, and three bullets hit my machine when I was flying an AH-1G Huey Cobra in South Vietnam. That was enough, I landed and told the cops no more stunts like that. The cops took all day to talk him out of the cabin, but by that time, I was long gone. The murderer was later judged crazy and was sent to a secure, funny farm for the rest of his life.

The Queen:

A week and a half later, Queen Fabiola, with her husband, the King of Denmark, showed up, and two helicopters were dispatched to take them to the Chateau Richelieu in Baie St Paul. A very large, very elegant hotel for the rich. It was a popular tourist place, much like a castle. It was very popular for tourists to come and watch the whales. To this day, there are several boats available for taking cruises to watch the whales.

On the way there, the Queen, who was sitting in the passenger seat beside me, saw piles of rocks on the sides of the many fields on the way. She asked if they were graves. No. I told her that the land grew rocks. In the winter, the ground froze deep, and the rocks were pushed up by the frozen water in the ground. In the spring, the farmers went around and picked them up from the fields and piled the rocks along the sides. Every field had fences of rocks piled on either side. A common feature in Québec on land that "grew" rocks.

The next day, when they finished their ceremonies, we flew back to Québec City. Queen Fabiola was a very beautiful and smart lady. I really enjoyed talking to her. What was really pleasant was that I had her all to myself on the flight to the hotel and then the next day back to Québec City. My machine had no intercom to include those seated in the back seat, so for over two hours, I had her all to myself.

New HQ:

The police got a new headquarters building, and it was great. Lots of parking, an MOT-certified heliport, lots of office space, a garage, and a firing range in the basement. They also had some cells, but they were one thing that I never visited.

I did visit the gun range fairly often. They had a stock of pistols, semi-automatics, and various machine guns. I got to fire any I wanted.

They found out that I was ex-military and had fired most of the weapons they had in stock except for the variety of machine guns they had. I got to fire each one of them and learned the characteristics of each one. I spread a lot of shell casings all over the floor.

In that, there is a story: Years before, cops were trained to put the shell casing from their pistols in a bucket placed at each firing position. Then, when they got into a real firefight, some stopped shooting because they were looking for the bucket to put their spent shell casings. They had reverted to training. The buckets disappeared fast after that incident which was during a real firefight somewhere in the USA.

Training sticks and really makes a difference in real life.

(Years later, at the Bell Helicopter training school I was teaching a student full-on autorotations. He was in his 50s and was a single pilot for a man who had his own company and was prosperous enough to afford his own personal helicopter. He convinced his boss to let him go for factory training in emergency procedures. He finished the one-week training program, both classroom and flight training. I was very happy with his flying performance. He was a good pilot; his problem was that he had not done any emergency procedure practice in almost 10 years and was very rusty in his procedures, so we did a lot of them. He went home on Friday and the next Monday, flew his boss somewhere to a meeting, and on his way back to base, he had a complete engine failure. He reverted to training. He completed a full-on, engine-out, zero-speed touchdown autorotation. His helicopter only required engine repair, and that is very expensive, but zero damage to the helicopter. Training works. Training saves lives. The pilot called our chief instructor pilot at the training school and told him what had happened and to let me know what had happened. He was a very happy student, saved his machine, and his boss was also very impressed with the value of professional training.)

I met the two snipers from the Emergence Squad and they invited me to meet them when they went out to practice their shooting in a large sand pit nearby.

They fired first, and they were very good. I owned several rifles, and I had started practicing shooting when I was only 14 years old, I had been taught in the military, so I knew how to shoot.

These two guys were the best I'd ever seen. One of them offered to let me shoot his rifle. I set up, took off the safety, aimed downrange, and put my finger on the trigger, getting ready to shoot. The rifle fired. I had never held a weapon with such a light trigger. He liked it that way. It took a few minutes to get used to it, but I really preferred a stronger trigger pull.

All in all, I did fairly good, but not up to their calibre of shooting.

The Bishop:

The Bishop of the Catholic church visited Québec City and was escorted around by the Provincial Police. He visited the new headquarters and after, was invited to go for a helicopter ride.

He was very reluctant, and I could see he was scared. I had to be extra careful with him. One of the problems with people when they get in a helicopter is that they want to look down. I couldn't let him do that. Right away after take-off, I pointed to the twin bridges on the horizon. I kept pointing out things for him to look at, always on the horizon. I told him to sit with his back straight and only tilt his head, not his body, when I turned and to always keep his eyes level with the horizon. After a few minutes, he wasn't even conscious of what he was doing; he just did it.

He did not get airsick at all; in fact, he loved it. After we landed, he was like a little kid with a new toy; he really enjoyed the ride.

Mission accomplished.

Flying 6th Sense:

We were called to a search in the Gaspé, 3 people lost for two days. By the time I got there the first day we didn't have much time to search. The next day was spent searching. They were not smart enough to build a smoky fire with which I would have found them in minutes.

I was passing low over the top of a hill when, looking out the right door, my head snapped back, and my eyes locked onto the three people between my feet so fast and so hard it hurt my neck. My 6th Sense really kicked in hard. That was about the third time my mind had seen lost people, but this time, it was a violent snap of my neck. I'll never forget it. After that, I let my 6th sense do its thing naturally. I always found the lost people after that if I got anywhere near them. My mind could see better than my eyes. It had taken over two years to develop, but it was working, finally.

That was not the last time I hurt my neck, but it was always satisfying. With the snap of my neck, I would yell, "Bingo!".

Just like my French, they both took time, and I had learned how to think in French. That permitted me to learn and understand French better and faster.

Anticosti Island:

Anticosti Island is a large, 100-mile, 160 km long island at the east end of the Gulf of St Lawrence. It had one town on it at the western tip of the island.

There was a report of a missing boat with six men aboard.

A search and rescue aircraft was sent from Halifax, Nova Scotia, to look for them. Their boat was found at the far eastern tip of Anticosti Island on the eastern tip of Heath Point. If I was to go there, I needed fuel, so the cops got the Canadian Coast Guard to send a ship to the eastern tip of the island.

I went to Havre St Pierre and left the next morning. It was over 60 miles, 100 km over water, to get to the island. I climbed to 9,000 feet to cross because if I had an engine failure, I would have a long time in the descent to scream for help. I knew the Coast Guard ship was less than 100 miles away waiting for me, so that helped.

The winds were picking up, and by the time I got to the east end of the island, the winds were 40 to 50 miles per hour from the east.

The east end of the island was packed thick with kelp. The kelp was thick enough to walk on. The men I brought along cut long, thin poles to explore the kelp for bodies. The huge waves born from the stiff winds caused the kelp to undulate, but there were no breaking waves on the kelp. I could see the Coast Guard ship only a mile off-shore waiting for me.

I cruised up and down the shoreline looking for bodies, but they were all in the kelp.

The probable cause of the overturned boat was the strong winds. Eastbound, the boat was heading into the wind but when they reached the point, they turned crosswind to go around to the north side of the island. The wind and the waves were too big and strong for the size of the boat, and it rolled over. The 6 men and the several deer they had poached were all in the kelp.

The news of this incident was kept very low-key because one of the drowned men poaching the deer was a police officer.

I searched the shoreline for a while, then went out to the Coast Guard ship to refuel.

The winds had not abated. The rough water was lifting the rear end of the ship where their helipad was located. It rose 6 to 8 feet with each wave as it passed under the ship.

I carefully hovered up to the ship with the landing crew ready. I was in radio communication with the ship, and that was good. It was all up to me now.

I hovered over the tail of the ship, watching the helipad undulate up and down. I waited a minute to catch the rhythm, then I put my floats a few inches above where the tail would rise, and as it came up, I touched down and quickly dropped the collective pitch to stay firmly planted on the deck.

Like flipping a light switch, I went from the motion of the air to the motion of the ship. Instantly seasick. If I had thought about it before I left, I would have bought some Gravol or Pepto Bismal.

I really wanted to barf, but I had to hold it down, or I wouldn't be able to fly. The Captain invited me to the bridge for a cup of coffee. I hoped that would help.

It was the first time I had ever been on the bridge of a ship. It was really cool and very interesting.

I took my time with the coffee; I didn't want to get sick.

After the deck crew had finished refueling my helicopter, I went back to shore where, during the time I had been searching and refueling, the searchers had found three men and a couple of deer. They still hadn't found the cop yet. One of the men had gone into the bush and shot a deer for "camp meat."

I soon left to get back to Havre St Pierre before dark. The searchers and the bodies would come out by boat later.

Stripper:

I travelled a lot for short periods of time. I did not want to cheat on my wife, so I did not go to bars to try to pick up women.

That, however, did not mean that my hormones shut down when I left home. The cops trusted me enough to give me an unmarked police car. For them, it solved a problem; they didn't have to make like taxis to drive me around town in the evenings. And since they paid my taxi expenses anyway, it just made good sense to give me a car to go to hotels, restaurants, and, in the evening, to local strip clubs for a beer and to rinse my eyes. In Québec, the women can strip down to skin, hair, and teeth. The most they would be wearing would be a wrist bracelet or earrings when they finished taking their clothes off. Most shaved between their legs, but not all.

They did their pole dancing, switching girls about every 15 to 20 minutes after they got naked, and one of their favourite songs at that time was "Betty Davis Eyes." That song I heard all over the province. The strippers were part of an organized group, and they rotated all around the province to do their routines in different towns.

One memorable young lady had a small bell, like a pierced earring, on her vaginal lip. (That must have hurt.) She would get guys to come up on stage, lay down, and she would bounce the bell on their noses. She then enticed guys to come up on stage and put a toonie on their lips. (Toonie in Canada was a $2.00 coin. Canada also has a loonie, which is a $1.00 coin with the carving of a bird, the loon on it. No more paper $1.00 or $2.00 bills in Canada are used.) She would then pick the toonie from their lips with her vaginal muscles, letting the bell bounce off their noses. There was never a lack of volunteers. The waitresses all kept a supply of toonies with them. The dancer could easily pick up 10 to 15 toonies in that same amount of time.

One memorable dancer I saw near Rimouski came into town, rented the largest conference or large dining room she could find, left posters giving time and dates, and charged admission. She set up her shows in all the large towns east of Montreal.

The rooms were packed, so I went to one. A stripper, but she put on a show for well over one hour. She did two shows a night for two or three nights. Her show was all about coffins and the spooky, and in the end, like all strippers, she was naked. It was very well done, with lots of props for her assistant to arrange during her show.

I later saw her on a TV news program in Québec City, where she talked about her shows during the interview, with only photos of some of her props, not of her performing her show.

Shots fired:

There are some idiots that think they can outrun a helicopter. Some have tried on the 4-lane highways, but they can't outrun a radio. My observer and I were well east of Québec City in St Anne du Beaupre. There had been complaints of snowmobiles running down several of the streets.

When we got there, we saw a snowmobile doing just that. We called in a police car and tried to stop the speeding snowmobile. Not only was he racing up the streets, but he was breaking any posted speed limits with complete disdain for any vehicles or children.

We guided a police car to apprehend the suspect, and he immediately got on the railroad tracks and used them as his trail. He then turned north to cross a couple of fields so he could escape into the trees. I got well ahead of him and landed across the trail. He then turned towards another grove of threes. He was well away from the road, so the cop car could not help us.

I again landed in front of him. He then took off in another direction. By then, my police observer was tired of this cat-and-mouse game, jumped out of the helicopter onto the snowmobile trail, and fired two shots from his pistol into the snow beside the trail.

The snowmobiler slammed on the brakes and jumped off his machine into the snow, and with his legs braced and his arms held high, he froze.

My observer got back in the helicopter, and I flew him close to the snowmobiler. With handcuffs on, he was put in the back seat, and we flew him to the waiting police car. After paying his fines he would have a long walk back to get his abandoned snowmobile.

What happened to him, I never knew, but complaints about snowmobiles in the streets quickly died in that town.

Incubator:

East of Québec City, there are a couple of large islands that only have ferry service, boats, or snowmobile access in the winter.

A young baby was sick and in distress. The local nurse on the island had the baby in an incubator with an oxygen bottle. They could not cross to the mainland in a boat with the incubator, the available oxygen bottle was too small to last the trip. I went to get them on the island. The incubator and oxygen were loaded on the back seat with the nurse in front.

On the way back to Québec City the incubator ran out of oxygen with only the small bottle that was available. The nurse turned around, opened the incubator, and, using a breathing bag on the infant, was pumping air into the failing infant's lungs. I called ahead and had the closest hospital to us prepare to receive us with a stretcher for the incubator and a larger bottle of oxygen for the baby.

It was just after dark when I landed, and there were too many people. Everyone wanted to help.

I had to jump out with the engine still running and the controls locked down. I ran around to open the left rear stretcher doors and guided the orderlies with the stretcher and oxygen bottle to the left side of the helicopter. One person had gone to the right side of the machine and decided everything was happening on the left side, so he started to go around the back. Not seeing the tail rotor in the darkness and obviously totally ignorant of helicopters, he was heading to his death. I caught him by the shoulders about five feet from the spinning tail rotor, lifted him bodily, spun him around, and literally threw him towards the front of the helicopter. I then grabbed him again, spun him around, and screamed in his ear about walking into the tail rotor and getting dead. I again roughly shoved him away and far forward of the helicopter.

By this time, the baby was in the incubator with the attached oxygen bottle and ready to be rolled into the hospital on a wheeled stretcher. I guided them clear of my machine, making sure nobody else was near the helicopter. I closed the doors, got back in the helicopter, and flew back to base. I was shaking all the way back to base. That damn idiot had just come too close to death, and I could imagine the huge uproar that a person getting killed by the Provincial Police helicopter would have produced.

(I later met a man in La Tuque who had walked into a helicopter tail rotor. He had done the same thing: walked around the back of the helicopter and walked into the tail rotor. The tail rotor blade hit him just above the right eye. He lived through the experience, but he had a dent in his skull just above his right eye that could hold a chicken egg or a golf ball.)

My son:

The year started quite busy. A lot of flying and many searches for lost people. One search was for 10 people who had not arrived when and where they were supposed to. A snowstorm had passed through that lasted two days and dumped a lot of snow. Snowmobile trails and roads in the bush were all covered with snow, leaving few trails identifiable. I was told the planned route of the group of skiers, and we finally found them off-route and still two days short of their next call-in point. We guided them back to a nearby lake and, the next day, met them with fresh food and got them back on the right trail. They were all fit and healthy, just a couple of days later than the worst guesstimation of their check-in point.

Then the Québec Carnival happened which was still enjoyable.

I again took my pregnant wife up with us over the nighttime carnival parade to give our unborn son his first helicopter ride.

There were canoe races on the river ice.

The first week of May was good weather, and I was in the backyard chatting with my wife in the bright sunlight when on my wife I saw big tears slowly dripping down from under her sunglasses.

I gently helped her to the car and drove her to the hospital, where, several hours later, my son was born.

When I got back home, I called mom-in-law. I would pick her up the next day at about noon.

When I brought my wife and newborn son back home, we sat my three-year-old daughter on the sofa and put her new brother in her arms. With the baby, I had bought a nice doll which we told our daughter was a gift to her from her new brother. After that, she always wanted to help change or feed her baby brother. Even with young children, first impressions are important.

Torngats:

The Bell 206B that I had been flying was due for a 1,000-hour maintenance inspection. It was to take several weeks. That left me with no helicopter, so the machine from Montreal would cover Québec City's needs.

I couldn't go anywhere on vacation with a newborn at home and with mom-in-law still with my wife, so all I could do was rattle around the house for a few weeks.

The Air Service solved that. The game and fish department had a job they wanted done. They sent me on a mission up north near Schefferville for two weeks. The job was to film caribou in the Torngat Mountains. They occupy the northern point of the Québec/Labrador Peninsula between the George River and the Labrador Sea, which is iceberg alley between Labrador and Greenland.

On my way up north, I had to stop for fuel. Between two fuel stops on one of my legs of flight, I saw the biggest moose that I had ever seen, and over the years, I had seen a lot of moose. I went low and circled around him. He was not one bit scared of the helicopter. In fact, when I got too low, he rushed towards me to attack. The moose was in a very isolated area of the bush with no roads for over 20 miles. I would suppose he probably lived to a ripe old age, but what a rack of antlers. A Boone and Crockett trophy, for sure.

At Schefferville, I met the film crew, and we went to the George River camp.

We set up our base far to the east of Schefferville at an outfitters camp along the George River. The camp was at a major crossing point of the George River for the caribou. At this ford, they crossed by the thousands. For the hunters during hunting season, they could select which caribou they wanted to shoot.

The camp was supplied with food and fuel by a De Haviland Otter aircraft. The Otter pilot was kept very busy supplying several camps along the George River. He always had people or caribou meat to take back to Schefferville after bringing in people and supplies.

The photographer wanted to film "the drop" from the mother caribou giving birth to a newborn caribou. He got legs sticking out the back but never "the drop".

He got me to leave him and his camping gear about 20 miles into the mountains that were all still snow-covered in June.

I carefully marked his position on my maps. The rolling, treeless, snow-covered terrain was my real challenge.

When I got back to camp, I organized my maps. There were dozens of lakes but they were all still frozen and snow-covered. The treeless terrain was all snow-covered and there were no land features that I could use as reference points.

My compass had over fifteen degrees of variation to cope with to navigate.

The only thing I could use was land elevation. I had to fly following the contour lines on the maps. That meant I had to fly at ten to twenty feet above the ground because any higher and the land was featureless. I followed the contour lines over what were small lakes and little hills. I followed creeks, skipped over hills, and used the contour lines until I saw him.

It was the most challenging navigational problem I had ever encountered.

I found the photographer, and he still didn't get "the drop," but he did get an interesting shot. He had climbed a ridge, and about a mile away, a wolf was trying to attack a newborn caribou. The photographer was kicking himself for having left his longest lens back in his tent.

The shot he did get was the caribou mother defending the newborn calf against the hungry wolf. The mother kept her head low so the wolf would have to get past the antlers. Several times, the photographer caught the wolf attacking, and the mother would hook him with her antlers and throw the wolf over her back. During this time the newborn caribou was struggling to get to its feet. The wolf finally gave up when the calf was on its feet. With the wolf trotting away in defeat, the caribou mother led her calf back to the herd.

The following days the photographer filmed many thousands of caribou.

A large herd was trotting past the camp to the ford in the river. We got a lot of film of them jumping into the water, swimming across the river, and many close-ups from the helicopter.

The two weeks passed quickly, and I flew over seventy hours in two weeks.

I left with the photographer and his precious film and the director with our personal gear for Schefferville when we were finished. The others came back in the Otter.

On the way back, I was going to Sept Isles (Seven Islands) for fuel, and there was a range of high hills just north of Sept Isles. I had to climb to 3,500 feet to get over them. The air was clear with a high overcast cloud layer. Suddenly, my windshield was a sheet of ice. Virga! It took about two seconds, and I was blind in front. I could only see out the side windows. There was the Moisie

River a couple of miles and 3,300 feet lower to my left. The upper level contained warmer air and it was raining from a clear sky. There was a front moving in, and the warmer air was riding over the denser, colder air. When I hit the rain, it had cooled to below freezing and it was these supercooled water droplets that turned instantly into ice upon hitting my helicopter.

The air temp was minus 9° Celsius, or 16° Fahrenheit.

Temperature decreases by 3 °F or 2 °C per thousand feet of increase in altitude.

I turned left toward the river valley and dove towards the river. My greatest problem was the rotor blades icing up and the intake of the engine freezing over. Right away, I turned on my engine intake and pitot heaters. The engine intake is heated by hot bleed air from the engine compressor stage. The pitot is the little pipe in the front of the nose where the forward speed pressurizes a diaphragm in my airspeed indicator, giving me my airspeed.

As I got lower near the river, the increasing temperature at the lower altitude started melting the ice from my windshield. The river valley was warmer than standard and I knew the air temperature on the ground at the Sept Isles airport from the ATIS broadcast. (Automatic Terminal Information Service). Now, I could see forward again.

From the camp back to Québec City, it took me over eight hours of flying.

Several months later I got invited to watch the caribou film. It was very good, well-edited, and the highlight was the distant wolf attack.

Blue Sea Lake:

I was sent up to Blue Sea Lake south of Mont Laurier.

A tornado had touched down along the shore of the lake. The destruction was massive for such a small place. Not near the size and destruction of an Oklahoma twister but it was significant for Québec.

The usual destruction of cottages, boats, docks, and lots of trees. A lot of photos.

Fortunately, no one was killed or badly hurt.

Buggered bug:

I did a lot of traffic, and we gave a lot of tickets using the helicopter, but this occasion was hilarious.

North of Québec City on the then two-lane highway, we were patrolling for tickets. We had two police cars on the northbound side to apprehend any lawbreakers. A Volkswagen was slowed down by a lineup of traffic, and the driver decided to pass on the right-hand shoulder. He got past the blockage (someone turning left, holding for the oncoming traffic) and got back onto the asphalt. Before getting to the waiting cop car, he, too, turned left. We sent the cop car after him. A couple of minutes after the stop, the cop called and told us we should come down and have a look at this car.

All four tires were bald, with no tread, and down to the cords. Why he didn't get a flat on the gravel shoulder was pure luck. He had household gutters on either side simulating header exhaust. (His Volkswagen engine was in the back). One-inch wide strapping with the holes in it was screwed onto his fenders holding large rear-view mirrors. He had screwed a windshield glare-shield over the windshield to simulate what trucks use. His radio antenna was a straightened metal clothes hanger. The driver's side seat was patched and taped. The gearshift was a piece of pipe bolted to the top of the transmission. His turn signal handle was a long screwdriver stuck into the steering column, and two of his dash switches were short screwdrivers.

There was other stuff done as well, but that's what I remember.

The cop was laughing as he was writing the ticket. He then called a tow truck. The life of that buggered bug was over.

Six days:

The Montreal machine was down for maintenance so I was called to go to Val D'Or (valley of Gold) well north of Montreal to search for a lost hunter. Québec City now had two dog masters assigned to us. One dog master had a German Shepherd, and the other had a bloodhound. I picked up the dog master and his bloodhound and went up to Val D'Or.

The man had been lost for five days, and we arrived on the 6th day. I picked up an observer from Val D'Or, and the dog master and his bloodhound went into the back seat. The dog master always carried his walkie-talkie and two spare batteries in his backpack. We went to where the lost man had started. The lost hunter had his rifle, a .303, a hatchet, and enough food for a day.

We toured the area where he was supposed to do his hunting.

We were scouting the area, looking for him or anything that he may have left behind. I started circling a small grove of trees with swamps on two sides, and among the fallen leaves, I saw something. I didn't know what it was, and I didn't know what had caught my eye, or maybe, in this case, my 6th sense. I circled for a few minutes and then told the dog master I wanted to drop him off to see what he could find. He walked into the spot that I found curious, and he told me the place was littered with fresh wood chips from someone using a small axe or hatchet. He then found a firepit that had been covered over.

Now, we had a starting point for our search. We followed the dog's nose for a few hours, with me ranging ahead of the dog, always working forty-five left and right of the dog's direction. The human scent was not where the man had walked, but off to the downwind side in the bush parallel to his direction. It made for rough going for the dog master, but it gave me a direction. We followed his scent for a couple of miles, and then I spotted a small tendril of smoke in a swampy area ahead. He had built a small fire on top of a small tree stump in the swamp. A place that was impossible to land, but I thought I could at least hover low enough for him to get in the back. There was no place close to land, so I got on the loudspeaker and told him to stay there. I called the dog master and told him I was coming to get him. I then brought him to a place as close to the now-found hunter that I could land. I asked the observer to stay there with the dog master. They had a radio and a map and knew where they were. I told them that it was a very dangerous place for me to get the hunter and I didn't want them with me. I showed them where he was waiting.

I dropped them off together and went to get the hunter. I hovered down into the scrub trees as low as I could go. He was able to climb onto the float and then into the back seat. I pulled up collective and, getting airborne and told him to give me his rifle. He still had a round in the chamber. I cleared the rifle of ammunition while flying back to the two cops and gave the hunter the half quart of milk and an apple that I had with me. It was his first food in five days.

In all the time we were there we never saw a moose.

After that, it was back to Val D'Or to put him in a hospital. The cops kept the rifle until he got out of the hospital.

The Net:

A few days later, I was called to go to Chibougamau, four hundred miles north of Montreal. One of the natives had left with his wife and two kids in a boat to go up the lake. After two days, he didn't return with his family, and the police were called.

I went up the lake, and in a small cove, I saw the torn-apart boat and the four bodies floating. We went back to the Indian settlement and sent a group of Indians up there to recover the bodies. It turned out the man had set a net to catch fish, and when he went up to clean out the net, he ran over it at the full speed of his boat. The net was strong and firmly anchored. The net tore the transom, the rear of the boat, and the motor right off the boat. None of them were wearing life vests.

Numbers:

It was the fall, and the new cars were coming off the production lines. New police cars were also being delivered. All the new ones had the vehicle numbers painted on the sides and back. In addition, the numbers were painted on the roof of the vehicles in two-foot-high and three-inch-wide numbers.

That was great, now we could call up and identify an individual vehicle from the air. I'd been waiting for that for two years.

Jack:

My wife and I started spending weekends with Mike, Pete, Jack, and other friends at Jacks barn/cottage.

Jack had bought a small farm with a bush in the back half and a large barn behind several houses over a hundred yards from the highway going through a small town south of Québec City.

Jack was a municipal policeman with a Midas touch. Everything he did, he made money. He knew all the court bailiffs, and whenever there was an auction or a bankruptcy, he knew about it. He would buy loads of stuff at giveaway prices and then sell them piece by piece. Mike's wife had a beauty shop in the basement, and Jack sold her a lot of stuff, which she resold to her many customers.

Jack also had his hand in warrants. Some people would get parking tickets and never pay them. The city wanted their money and Jack would wait until a person had five to ten tickets, then go find them. The parking ticket was $20.00 to $25.00. By the time it had been processed for non-payment, the tickets given by Québec City had risen to over $130.00 each. With the license plate numbers and his cop connection, he could find anybody. I knew about this but had nothing to do with it in the beginning.

Jack bought an Excalibur, a really fancy, fast, 1930's looking car. A powerful GM powertrain and two tires mounted on either side of the engine, a black box behind the four-seat interior as his trunk, with a 12-inch high windshield with 3 wiper blades. It also had four bug-eyed headlights and chrome headers coming from either side of the engine hood that were real, all on a 125-inch wheelbase, the same as a Rolls Royce.

There were times when we would be invited by Jack and his woman to go out to a fancy club for a few drinks. He would pick my wife and I up at our place and go to the club, where he would park the

Excalibur right in front of the front door of the club, in a no-parking zone. Nobody would touch it, but everyone would look at it. It was the only Excalibur in Québec City. When we came out of the club there was a $25.00 parking ticket under the windshield. Jack told us it was part of the cost of going out and parking in a secure location. We split the ticket cost.

Grey Camero:

The year started fairly quiet, with a few searches, a lot of tickets, and aerial photos.

Then, something different. The cops wanted to follow someone. A crook, they wanted to know where he went and to see whom.

He always thought the cops were following him, and he took elaborate measures to shake them off his tail.

Well, he didn't know about helicopters.

His car was identified to me by the cops, and I climbed to 5,000 feet behind him. He was never out of my sight all the way to Three Rivers (Trois Rivières). He doubled back, took side roads, and parked for minutes at a time along the highway. Did everything he could to try and see someone tailing him, but he never got out of his car and looked above and behind him.

I followed him to his destination, and the cops were really happy because it was the first time they had identified his destination. Upon arrival at his destination, he hid his car in a garage and thought he was safe.

The cops got out some long lenses for their cameras and were happy with the chase because they got to identify his accomplice. The aerial surveillance worked.

Cop killers:

A Police officer north of Three Rivers pulled over a car that was speeding. He called in the license number and then went to get the driver's license, insurance, and registration from the driver.

The driver shot the police officer dead, then sped off and went into hiding. The following vehicles saw the dead police officer and reported it.

A massive manhunt went underway. I arrived with my observer, who was the shooting instructor for the Three Rivers police headquarters. We started looking at every possible place they could have hidden. The place was up in cottage country, and there were dozens of places he could have holed up. There were reported to be three suspects in the vehicle we were hunting.

There were very few good places to land, and the police wanted to spread out their forces for the search.

I knew the shooting instructor. He carried a .44 Magnum revolver. I suggested that he let me use a rifle to cover him when he got out of the helicopter to search the cottages. I told him I wanted a rifle with open sights because if I saw someone pointing a gun at him, the half-second it would take me to line up the murderer through the telescope was too slow. He gave me a .30-30 lever action Winchester. I had one very much like it; in fact, I had killed my first deer with my .30-30.

I told him that, plus my military experience, plus that I had gone shooting with the snipers from the emergency squad officers in Québec City.

Thereafter every time I landed and he got out to check the area, I got out of the helicopter with the controls frictioned down, and with the engine at idle, I covered his back.

I never let the muzzle of my rifle cross him, and I always kept the muzzle pointed behind him to get me the best coverage of his back. I knew that if one of the suspects popped up behind him for a shot, I would have to react fast to get the first shot off at him, so I had the safety off and my finger on the trigger. Even if I missed the suspect, which I thought unlikely, the shot would warn the police officer.

We continued this for the whole afternoon and again the next day, covering dozens of places where the cop's murderers could be hiding.

The heat was on them, and they knew it. They stole another car and were attempting to leave the area the next night when they were stopped by a police barricade, and with several shotguns pointed at them, they gave up.

The police had set two cops in a car well ahead of the barricade, and they advised the cops at the barricade about any vehicle that had three men in it. It worked.

They all spent a long time in prison.

Call two CL 215's:

I went up to Roberval north of Québec City, and about 100 miles north of Lake St Jean a fisherman had been lost for two days. The cops knew where the person's pick-up truck was, but it was a lonely area with very few cabins.

I had arranged to have a truck bring up some fuel.

I started the search and tried to stay within five miles of the truck. Nothing, but the lost person would have heard the helicopter.

I stopped for fuel, and when I had refueled, I started searching in a different area. I saw a thin tendril of smoke a few miles away. I located the man and saw a small lake about half a mile from the man. It was an area of dense bush with no place to land except on a small lake. I dropped my observer off with a walkie-talkie portable radio and guided him to the lost fisherman.

When he met up with the cop, I noticed that they headed back towards the lake where I could pick them up, but they didn't put out the fire.

I had to climb up to over 6,000 feet to get a line of sight over the hills to get Roberval radio. I told him to send out the Air Service CL 215 water bombers to put out the little fire before it grew to become a big forest fire.

I also told him to tell the water bomber pilots to load up with water before they got here because there were no lakes large enough for them to load up that were close to the fire.

That was the only time I ever had to call water bombers after a lost person fire.

On my way back south to Roberval, I passed 2 CL 215 water bombers with their bird dog aircraft racing north to the fire.

Blade hook:

Benoit, the chief pilot, was out on a mission with the Game and Fish Department people to net some fish they had put in a lake two years before. His helicopter was on fixed floats, so Benoit landed on the water, keeping the helicopter at flat pitch and 100% operating RPM. The game warden got out on the float with his grappling hook to snag the net. But when the Conservation office threw the hook from the helicopter towards the net, he threw it too high. The hook hit the rotating blade and threw the hook back over the tail boom of the helicopter, fortunately missing the tail rotor. Luckily for the officer, he had not wrapped the rope around his wrist and the rope was just pulled from his hand.

Ben then landed on solid ground and shut down to inspect the damage to the blade.

He had a good-sized hole in the blade from the tang of the treble hook.

It scared him. He restarted the helicopter and climbed high enough for the radio to reach a line of sight with the antenna on the high hill to the north of the Québec Airport. He described the incident to the distant dispatcher and went back to the lakeside and landed.

The dispatcher called the police for permission to use me and their helicopter to go and get Benoit. The Police Emergency Squad dispatcher immediately agreed.

The Air Service dispatcher described the problem Benoit had to me. I told a mechanic to bring a ball peen hammer, a roll of duct tape, and a step ladder with us in the police helicopter to help Benoit.

When we got to Benoit, I had a look at the blade and told him it was just a bit bigger than a .50 calibre bullet hole. It did not damage the blade spar.

I had seen that in Vietnam and knew what to do.

I took the ball peen hammer and tapped the rough edges back into the holes on top and below the blade. I then carefully wrapped a couple of turns of duct tape around the blade to cover the holes with the tape ends cut at the trailing edge of the rotor blade. This action prevented the tape from being peeled off by the force of the air on the rotating blade.

That done, we both started up our machines and flew back to Québec City.

Benoit was amazed. He had no vibration problem at all with the fix.

A new blade had to replace the damaged blade, but that was done in the hanger, not in the bush beside a lake that Ben had envisioned.

BC motorcycle:

It was a busy summer, with many searches, a lot of traffic, and pot hunts, and they kept me busy all around the province.

On my way to Baie Comeau, the wind, for a change, was quite calm. Just before Tadoussac, there was a thin layer of fog that was only a couple of feet above the water. The noise of the helicopter had scared up a couple of seagulls, and one had flown over the layer of fog by only a few inches ,with each flap of its wings, left two holes in the fog from the wing tip vortices coming off the tips of its wings. It made like a series of parallel dots in the fog. I circled around and, from lower down took some photos of this first seen phenomenon.

"Tracks of a bird in flight" was the name of the photo in a photo contest.

(That reminded me of another happening with a bird. I was flying in Labrador and very low along a river. The trees were fairly high on the sides, making a corridor of the river. I scared into flight several Canada geese. They pealed off until only one was left. I was getting very close to it, and the goose tried to roll out of my way. It snap-rolled right and stalled. I had time to see it roll violently to the left into the stalled left wing, and as I passed over the goose, it hit the water. I rolled right and kicked the right pedal in time to see the ruffled, but unhurt, goose start its take-off back the way it came.

First time in my life that I had ever even heard of a bird stalling. (To have seen it happen was incredible.)

In that same area in central Labrador, I was once eastbound down a valley with the sun behind me and a light rainstorm in front of me. The effect was a series of 3 perfectly round, complete rainbows. The center one was the brightest and the outer one the faintest, but all 3 were spectacular. The effect lasted only a few minutes until I flew into the rainstorm. I really regret not having a camera with me.

Before going to fly in Labrador, I was afraid of low-flying jets in that area. Goose Bay, Labrador, and Gander, Newfoundland, both had large air force bases. Canadian, American, and several European NATO allies sent their fighter pilots there for low-level flight training. I flew at a low level. There are 2 types of notices for flights that relate to me. A "Flight Plan" which is a point-to-point advisory of a flight with times and has flight routes marked for search and rescue, if needed. A "Flight Plan" is opened upon departure at the departure point and closed with an Air Traffic Control authority upon termination of the flight. The second was a "Flight Notification" that gave dates and areas of the flight advisory. I gave the coordinates of the 4 corners of the area that I would be using for the next 6 weeks and filed that on my "Flight Notification." I never saw a single low-level fighter jet in the 6 weeks that I was in that area. I closed the "Flight Notification" when I left the area with a call to the Goose Bay control tower when I went there for fuel on my way home.)

In Baie Comeau, we were given tickets, and one guy tried to get away. That was rare. This time, it was a motorcycle. The cop with a radar caught him speeding and tried to get him to stop. The speeding motorcyclist just ignored the cop and soon lost the cop in the hilly terrain. He didn't know about the helicopter. About 10 miles further was a small bridge, and we called ahead, and two cop cars blocked the highway. They got out their shotguns and waited for the motorcycle. When he saw the cops, he slowed down and stopped. I then went low and buzzed over him. With two shotguns up his nose, he got really respectful. I soon heard the cops calling for a tow truck.

The motorcyclist paid a lot of money for his little speed run and ignoring the cops.

Cadillac Man:

A few months later, a similar incident happened.

. We were patrolling the east end of Québec City, getting near Orleans Island and we had a Corvette stopped for speeding. The police officer was beside the Corvette when we called him to stop a speeding Cadillac. He tried, but the Cadillac driver thought he could get away from the cop, so he floored it and took the next exit while the cop was still getting in his patrol car. After the exit, the Cadillac got to some residential streets. Here, there were several stop signs. He blew through five of them. In the meantime, the patrol car we were working with and another one that was close by followed our directions to apprehend the Cadillac. When he was stopped, I passed low over the top of him and saw the cop pointing up at us with the dejected driver leaning out his window for a good look.

While one patrol car held the Cadillac, the other one went back to the highway exit, and my observer talked the cop car through the route the Cadillac had taken. He added up all the stop signs and passed that to the cop holding the Cadillac. The first cop then went back to finish with the waiting Corvette. He could not have gone anywhere; the cop had his driver's license, registration, and insurance papers.

Little did I know I was also to find out the sequel to this Cadillac man's problems.

My friend and neighbour were an insurance salesman.

I just happened to tell Mike about the Cadillac and asked how that could affect his insurance.

A couple of months later, he told me what had happened to the Cadillac man.

First off, he was very rich. His father owned a very profitable stone quarry.

He went to court and lost his license on demerit points. Just the stop signs at two points each was ten demerit points, plus the speed, the failure to obey a police officer, etc.

When his customer, the Cadillac man, came to Mike to renew his insurance at a greatly increased rate, he told Mike what had happened. Mike already knew, and he kept his mouth shut. Mike did not tell him that his friend and neighbour was the pilot of the police helicopter.

His total was 17 demerit points, plus the fines and a one-year suspension of his driver's license. He had to hire a driver to go anywhere he wanted for a year and then get his license back, going through the same requirements as any 16-year-old.

Driving courses, written tests, beginner's permits, and very high insurance rates.

The Cadillac Man told Mike he didn't give much of a damn about the money. It was the demerit points that really hurt.

"Jesus" nut:

I had been to Montreal, and on my way back, I stopped and picked up a photographer. In a large field beside the Beloeil Airport, a small airport east of Montreal, a Bell 206B helicopter and crashed into a very muddy field partly covered with water. A tall wheeled tractor was needed to get to the crashed machine. The rotor blades and attached hub were several hundred yards from the downed helicopter. The blades and rotor hub flew off, and the helicopter turned into a lawn dart, killing the pilot and the mechanic.

Back in the hangar, the investigators were shown the Main Rotor Retention Nut, otherwise called the "Jesus" nut, on top of the mechanic's toolbox.

The mechanic didn't check his work.

The mandatory inspection required when a flight control is disconnected or removed was not performed, and:

The pilot did not do the required pre-flight inspection.

Two dead from their own carelessness.

Roots

Flash-bang:

The Québec carnival parade started as usual. The cops with me kept things down below organized, and when the end of the parade reached the Parliament buildings, the fireworks started. The first launched was a flash-bang. Very bright and very loud, like a ¼ stick of dynamite.

I was passing in front of the parliament building when the flash bang went off about 50 feet from my left door.

Not only did it just about scare the shit out of me, but thereafter I stayed much further away when the parade ended. I learned.

Dynamite:

The cops wanted to test the effects of dynamite on ice for whatever reason unknown to me.

They had the game and fish department tell them of a lake in the hills that had no fish in it.

I put the dynamite and some det cord in the cargo compartment, and the cops carried a small metal box with the detonators in it in the front seat where I could see them.

I had shot almost every kind of rifle or cannon up to 20mm, and expended a few tons of rockets, grenades, and bullets, but I had never blown anything up. There's a first time for everything.

They told me to blow every charge they set up. A small amount of C4 could blow a fairly big hole in 12 inches of ice, and a couple turns of det cord could bring down a good-sized tree. Det cord or detonation cord is like a piece of plastic-coated clothesline, only inside the plastic was not steel it was C4 explosive. C4 is a very safe explosive to use. It's like silly putty and can be easily moulded into any shape. It requires a detonator, a smaller explosion to set it off. (A Composition C 4 is an RDX {Royal Demolition Explosive} and a PETN- based detonating cord.)

Broken neck:

On a clear, cold winter night with a snowstorm on the way, I got a call at about 21:00 asking me to go get a man with a broken neck up in the hills.

I called the chief pilot and asked him. An emergency situation he told me, your call. Up to me.

For this, I needed Frank, or I couldn't go. I called and told him what the problem was and where I had to go. He knew exactly where I needed to go.

The place was in the territory he usually patrolled as a volunteer game warden.

Frank got his snowmobile and went to the gate to enter the area where the trail started to go up to the lake where I would have to land. A cop was waiting for him and the friend of the injured man.

Damn, politics reared its ugly head. The friend of the injured man was a cabinet minister of the present government, knew and did not like Frank. He refused him permission to help. I was contacted and told the cop that if Frank doesn't go, I don't go. He can tell the politician his injured friend can either die up in the hills, get on a sled behind a snowmobile, and die from the buffeting he would receive on that, or Frank goes up to help me.

He didn't like it, but he got the hell out of our way. I told the cop to give Frank a handful of highway flares.

I then told Frank exactly what I wanted him to do.

In Canada to fly a machine with a passenger on board at night required two engines. My helicopter only had one. I had a night rating and had flown at night every year for the carnival, but at night in the hills with no references was out of the ordinary and very dangerous.

I called dispatch and got a doctor to go with me. They always had one on call for the air ambulance missions they often flew.

I left the Air Service with the doctor aboard and told the air service to put an ambulance on standby at the Air Service hanger.

Once I got to the hills, my only visual reference was the stars. No lights in the hills.

Frank heard me coming and saw my navigation lights. He stood about 200 feet from the edge of the lake straight out from the small cabin where the injured man was lying in a bed in the cabin. A wood fire had been started in the stove, so the cabin was at least warm.

Frank popped two highway flares and then got to his knees on the snow of the snow-covered lake. Frank was my reference point. I was going to fly myself into a whiteout condition at night near the middle of a small lake. It would be like flying into a ping pong ball at night. No references, which is why I needed Frank and the flares.

I made my approach directly to him and landed with the nose of my helicopter about two feet from his face.

We went into the cabin with the doctor and the helicopter's stretcher to get the injured man to bring him back to the helicopter on a sled.

We got the injured man to the helicopter with the doctor and then closed the doors. The take-off was the reverse of the landing. Frank was in front with two more highway flares and a vertical takeoff until I was clear of the whiteout.

Frank later told me that the injured man was the owner of several jewelry stores in Québec and other towns not far away. In other words, a very rich man.

The injured man was taken to a hospital, and I thought that was the end of the event.

No so.

Later that year, during the summer a newspaper reporter wanted to do a story about the police helicopter and the two pilots.

We told him the various missions that we had done and I mentioned the jewelry store owner with the broken neck that I rescued at night just before a snowstorm. I added that he was too cheap to buy a stamp and send us a thank you card.

The next Saturday that story with photos of us two pilots beside the police helicopter was on the front page of the newspaper, and with the story of a night rescue patient being too cheap to buy a stamp.

A week later, I got a furious letter from his wife, but she never did say thank you for saving her husband's life.

No good deed ever goes unpunished.

30-Z:

Frank and I later talked about the numbers on the top of the police cars. He wanted to do something so I could identify his pickup.

The police numbered their cars by district. Québec was district 3 and so all Québec district vehicles started with a 3. Emergency squad cars were 30- then the car number like 30-05, Québec metro vehicles were 31-01 to 31-99. All the sectors in the district used the second number as the sector number, and the last 2 numbers were the vehicle number.

Montreal was district 6, so emergency squad cars were, for example, 60-01.

My helicopter was 3h-01, which was never used. We used the aircraft registration, CF-PQV.

Montreal's helicopter was 6h-01.

Since I could not give Frank a police number to use, I told him to use 30-Z.

The next time I saw his pickup, the number 30-Z was on the roof, same size and colour as the police cars. Frank wanted to be part of the team and was making an effort to be just that, and available.

The General:

A young handicapped girl was lost north of Québec City. She was from a home on the north side of the Valcartier Military base. The girl was last seen heading south along the road that passes the military base entrance. We had a description of the girl but did not find her along that road. No other indicators were offered and we spent time searching the whole area outside the base. Valcartier is a base for military helicopters and had a control tower frequency for their area, but it was for a heliport, not an airport. I contacted them and told them that I would be in their airspace looking for a lost person.

A couple of weeks later, I got a letter from the Commanding General of the Valcartier base telling me that he was going to report my intrusion of his airspace to the Ministry of Transport and give my pilot's license violation, which could lead to a hefty fine or a license suspension.

I took the letter to the Director General of the Provincial Police headquarters in Québec City. In essence, I was working for him.

He told me to tell the General that I was looking for a lost person at the request of the Provincial Police. The highway was in Québec, the airspace was in Québec, and the helicopter was the property of the Provincial police and if the General was not satisfied, he was to call the Police Director General

182

at police headquarters. I included the name and phone number of the Director General.

When the General received the letter, he knew he was outgunned and I never heard another word from him.

Sometimes, there are people who, basking in their own self-importance, step over the line and have to be put in their place.

When the General made the complaint, he had to know he was dealing with the police helicopter.

Party Barn:

I spent more and more time with Jack at his cottage/barn. Mike, Pete, and I were usually there, along with several others, and it was always party time. Our kids were still young, and we would lay the baby on the big bed in the downstairs bedroom with everybody's coats forming a donut around the sleeping child. How he slept, I'll never know. On the other side of the wall were four large speakers belting out the dance music. The floor bounced along with the dozen or more dancers.

Getting serious, Jack usually had another cop with him when he went to collect his warrants. He decided to invite me to go along with him. He would pay me for every warrant he collected. Most of the tickets were parking tickets. With late fees, court costs, and administrative fees (mostly Jack's commission) the total per ticket could get up to nearly $100.00 each. Jack seldom went to find someone with less than 5 tickets owed. He got his commission and I for $2.00 per ticket collected. Never less than $10.00 a visit for collection, and we'd do 7 to 10 per evening.

My job was to stand behind him. We were both six-foot-tall and strong-built men. Nobody was allowed behind Jack except me guarding his back when he worked with the person he was collecting from. When they didn't have the cash on hand, he would force them to either go to a bank machine or borrow the money from someone else. Their alternative was a ride to prison, where they would spend the total time allotted on each ticket. He usually got paid. Some didn't care; they would go to prison, get three hots and a cot for a couple of weeks, and save the money.

Jack learned quickly which ones these were. He'd wait until 24 December and spend the day finding and collecting the money.

The prison would free a lot of their prisoners from the medium-security prison, and there was lots of room for the new arrivals on Christmas Eve, and they knew it and paid, not wanting to spend the holiday season in jail. Jack told me that he would sometimes have people talking in English when he was there. Jack couldn't speak any English, so he used me for that reason as well. Great, I made some free money.

One thing I used to do when I would drop off my observer for lunch and fly back to the airport for fuel and lunch myself, was pass over my house and give a quick chirp from the siren.

Mike used to hear me, and he told me that he told some of his customers who heard the siren that it was me letting my wife know that I would be there in a few minutes for lunch and that one chirp from the siren was for chicken soup or two chirps for pea soup with my lunch. The funny thing was, is that they actually believed him, and I don't like pea soup, and Mike knew it.

Liberace: (Liber-ott-chee)

Liberace was a gifted pianist and showman. His piano concerts were always packed with spectators, and his Grand piano always had a very large candelabra filled with a dozen or so burning candles. His trademark.

Jack asked me if I wanted to go to one of Liberace's concerts with my wife. Not just yes, hell yes. My wife was ecstatic.

So, the women got all dressed up for the concert. Jack and his woman and me and my wife all went to Montreal in the Excalibur.

We left in the mid-morning because Jack had a job to do before the concert.

A woman was working for a lawyer in Montreal who had a branch office in Québec City. The woman, his secretary, would frequently go to Québec City to pick up and deliver legal papers. (Fred Smith's Federal Express was in its infancy). She also got a lot of parking tickets in Québec City.

Jack had a pile of outstanding warrants that totalled just over $4,000.00.

We went to the lawyer's office to find the woman, and Jack placed her under arrest.

Of course, she panicked. She went to her boss and arranged a loan of money from the firm.

This took a few hours, and while we were inside their Montreal office, we were keeping eyes on the woman. She was scared. She had 2 large men guarding her, and prison loomed huge in her fearful imagination. We were waiting until the lawyer's office in Québec City got someone to go with a certified cheque to the Québec Municipal Police headquarters.

Meanwhile, our women were outside with the Excalibur. They were all dressed up to go out to the concert and the two beautiful women and the exotic car drew a very large number of spectators.

When Jack confirmed that the payment had been made, we left the devasted secretary behind, and we all went to the concert hall to listen to Liberace tickle the ivories. He was very good and put on a great show.

After that, we all went to a gourmet restaurant and had an excellent meal.

The entire trip, gas, food, wine, and tickets to the concert, did not cost us one dollar. The secretary paid for it all with our cut from her fines.

Aircraft and scuba:

I had to go to Chibougamau, 400 miles north of Montreal. A float plane had crashed, and the bodies were still in the cockpit. The weather in the park was again low clouds and rain. The high terrain in the park often invited the low clouds to proliferate, making the time longer and stressful. Dropping down the other side of the mountains, the clouds broke up, and visibility improved a lot.

Emergency squad scuba divers were sent up to Chibougamau with all their scuba gear. With full fuel, they were a heavy load with all their equipment.

The clear cold waters of the lake holding the 4 men were inside the aircraft and had to be taken out of the twisted wreckage. The divers worked from a small boat that had been brought to the lake.

The four bodies had to be taken out and were tied onto the floats of the helicopter two at a time. A vehicle was at the end of a bush road and had brought me fuel. It would also take out most of the scuba equipment, and I took what I could, plus the two divers. There was also an ambulance that would take out the bodies. The lake was quite a way from Chibougamau, and I was getting fuel-critical when I got back to the fuel pumps.

Back in Chibougamau, I was told that on my way back to Québec City, I had to stop at Chicoutimi. Five people were lost east of Chicoutimi, not far from the Saguenay River, about halfway to Tadoussac.

When I got there the next morning, I went to the district headquarters and met the sergeant who wanted to come with me.

We got into the helicopter, and I gave him the map. He showed me where he wanted to go and meet up with a police car on the main highway at the entrance to the road leading to where the five lost people were camped.

When he showed me the small lake where they were camped, I told him I knew where they were.

They would be at the same pond behind the small beaver dam where I had found the lone man who had climbed the hill a few years before.

He said no, he wanted to go to the police car.

I told him no. I did not come there to fly over a police car. We came to find five people, and I knew where they were from 50 miles away.

He was the sergeant; he was the boss and knew it. I told him he was wrong. I was the boss in the air, and it was the first time he had ever been in the helicopter. I again told him that I knew where the campers were. He didn't believe me. I told him this is your first time in a helicopter, and you may know things from the ground and be the boss there, but I told him I'd been a police pilot for several years now, and I knew what I was doing. We can talk to the car on the radio; there is no point in going to where he is, that's not where the people are lost.

He was pissed off at me.

I didn't care; he was wrong. Well, so was I, not to follow his wishes.

I flew straight to the lake and up the hill where I knew the pond to be. I rolled the helicopter left at about 200 feet above the five people beside the pond.

I circled around until I was about 200 feet above the campers and pointed down from where the dammed waters spilled around the beaver dam. I got ahead of them and turned my landing light on. I stuck my arm out the window and pointed towards their camp. That sent them down the hill to their camp. Fifteen minutes later, they were all at their campsite.

The sergeant had been on the radio and had called the police car to come to the camp and identify the five campers. When they joined up, I turned and headed back to Chicoutimi and police headquarters.

The sergeant said very little on the way back. He was pissed at me because I didn't obey him and also pissed that I was right. I think, I hope, he learned that the pilots were pilots because they knew what they were doing.

Obviously, I never saw him again.

Indicator Lake:

Indicator Lake was a long way north of Chibougamau.

I got there with an observer from Chibougamau and the dog master with his bloodhound with me from Québec City.

It was a prospector camp, and they had several natives with them for a total of about 12 people, plus their helicopter pilot and mechanic. They had cases of high DEET content OFF spray and lotion against the clouds of mosquitoes and black flies.

The camp boss was pissed, mostly with himself.

The camp cook was a native, and the camp boss wanted to finish the survey as quickly as possible and sent the camp cook out to assist in the survey. The cook was supposed to walk a straight cut line to another cut line and turn left, south, following another cut line, leave his markers, and then come back the way he came. Very simple, follow the cut line for a mile or so, turn 90 degrees left, follow another cut line for another mile or so, then turn around and come back the way he came.

The camp cook was a town Indian, not used to doing a job in the bush. He was the cook. He had no bush sense at all. He went a couple of miles, placed his markers, made his turn to place more markers for another couple of miles then came back.

With no bush sense what so ever, he tried coming back by cutting the hypotenuse. He had no compass and was so lacking in bush sense that he didn't even try to guide himself using the sun.

The camp helicopter couldn't find him, so they called on their HF radio to their base and requested help from the police. The police got me to bring up the dog after the second day.

I got as far as Chibougamau on my first day, and early the next morning, it was another hour north. The cook was now lost for 3 days.

It was late morning when we started the search. Like before, I stayed within 45 degrees of the dog's nose. It took me just under three hours to find him wandering in the bush. He had so sense of direction and had wandered a long way in the wrong direction. I picked up the dog master and took him to a small swamp about half a mile from the lost cook. It didn't take long; the cook rushed toward the sound of my helicopter idling in the swamp, and when the dog master met him, he brought the cook back to me in the helicopter.

I was amazed when I saw the cook. His face and hands looked like raw hamburger. He was not a very bright light. He did not spread mud on his hands and face to protect against the mosquitoes and black flies and had used up his single spray can of OFF mosquito repellant the first day. It was the perfect breeding ground for both bugs. Lots of small standing water pools for breeding mosquitoes and several small streams for breeding the black flies.

We got him back to camp, and I refueled and headed for Chibougamau.

I had never seen a person so thoroughly eaten by bugs, and I have spent a lot of time in the bush. It would take weeks for his skin to heal.

I got as far as Roberval. There was a front passing through, and the hills were socked in with solid low clouds, fog, and light rain. I refueled and continued on the Chicoutimi. Same front, same low, impenetrable clouds.

This time I took the hard way. From Chicoutimi, right down the Saguenay River, past the statue of Notre-Dame-du-Saguenay that I couldn't see up in the clouds, to Tadoussac, then turn right up the St Lawrence River to Québec. It, too, was almost socked with MVFR (Marginal Visual Flight Rules) conditions with cloud at 500 feet AGL. (Above Ground Level)

MVFR is flight conditions less than VFR (Visual Flight Rules), which is a minimum of 1,000 feet overcast and 3 miles visibility.

I spent the next couple of months doing searches, rescues, races, tickets, a suicide, and two murders. Routine, repetitive. Searches? I was getting used to getting a sore neck when I found someone, with the resultant happily exuberant "Bingo" to share my success with my observer.

Gold Fraud:

I met the pilot from Montreal with his helicopter in Chibougamau. On the way there I had no problems; the park weather was good visibility and high scattered clouds. Ed had brought with him two investigators with him from Montreal. The problem was that a woman from Montreal had complained to the police that she had been defrauded out of almost $100,000.00 in a gold mine scam. She was shown several pieces of quartz that were layered with gold, and the fraudsters needed money to develop the mine. After a while, she got suspicious and called the police. They studied the paperwork, the people backing the gold mine claims, and started an investigation.

The final step was to go to the mine and arrest the people involved.

That's where Ed and I came in with the two helicopters.

The cops wanted to surprise the people there and rush them before they could destroy any evidence.

I got Ed and the first assault wave of police officers together and told them we could do a Vietnam-style assault. I explained that to them.

I had been in Vietnam with the US Army and participated in many helicopter assaults, so I knew what I was talking about. They agreed.

The cops had maps and aerial photos taken from rented aircraft overflying their camp.

I told Ed that I would come in low and fast and then do a quick stop to a landing on the sandy beach. I told Ed to fly fast over top of me just as I was landing and make a tight turn over their heads, causing as much blade slap as he could.

He did exactly that. I landed, and three cops jumped out with their shotguns and charged into the men, staring at the two helicopters in frozen surprise. I lifted off, and Ed came in with his load of cops.

I circled around and, came back in, and landed not far from Ed's machine.

The cops had ordered everyone outside, and when Ed and I walked up, one of the suspects commented that it looked like an assault he had seen in Vietnam on TV.

I told him that I had flown gunships in Vietnam for a year and had been on many helicopter assaults. He went bug-eyed and speechless when I told him that.

The police found the complete written plans for the fraudulent scheme.

They also found a complete scheme for a fraudulent diamond mine.

I later learned that the son pleaded guilty, and the real planner, the father, was let go.

Like O. J. Simpson, get a good lawyer, and you can walk away.

Moses:

Then there was Moses.

Moses was a smart religious nut who set up a community in the Gaspé peninsula. He and his followers stayed out of the eyes of everyone except for the locals and a few government officials until one of his followers went to the police.

His followers told the police about the unreported deaths of a woman and her child. He also told the police that he was the father of the dead child and Moses had neutered him. He was still a young man and didn't like it.

The community of Moses was several miles into the bush on government land.

In one way, Moses was smart. He took advantage of every federal and provincial law or program that he could.

He found an old law almost 200 years old that obliged the government to give settlers on crown land setting up a community a horse for free. This old law obliged the government to give them a horse.

He got his horse.

He found other old laws and some that said the government would pay them welfare or a subsistence living allowance. Moses had the money from each of his followers deposited in the same bank account that he controlled.

With this money he could buy all the meat, tools, fuel, clothing, and a wagon for the horse to pull, and all the other essentials needed by his community.

He and his community were left alone and prospered for more than two years.

Then he broke the law.

He did not report the deaths of the mother and her child, and he performed a medical procedure on the body of someone against their wishes.

In any hospital in Canada or the US, the patient must, or a responsible family member must, sign a form authorizing the medical personnel to do the procedure before any surgery may be performed on the patient. That's the law. Non-compliance has very strict penalties.

The access for the police was by helicopter. The normal path going to the community was by horse-drawn wagon, or on foot.

I landed in the middle of their community, deep in the bush, with my observer and two investigators aboard.

While the cops were busy interviewing the community members, I was free to wander through all the buildings.

All were made of logs. The floors and the interior walls were all plywood and well insulated, plus good quality doors and windows were installed. They had money.

The storerooms were large and filled with canned and pickled food. Hundreds of mason jars lined the walls filled with preserved food, plus a dozen barrels of fuel for their oil lamps scattered throughout the buildings. They were certainly ready for the coming winter.

They heated with wood, and one of their main chores was cutting and splitting the wood, getting them ready for winter.

Moses was told he had best get a lawyer because he was charged with some serious crimes. The two bodies were later exhumed and taken to the morgue.

I found later out that he did go to jail for the deaths of the woman, her child, and the sexual mutilation of one of his community members.

With the departure of Moses his community members in the bush simply left. They had no more leader.

On the way back, we got diverted to Montmagny on the south shore. A passing boater reported to the police that he had seen a body on the shore of one of the small islands east of Orleans Island.

I picked up a body bag and two cops in Montmagny.

The body was hugely bloated from being in the water for so long. It was easily well over 400 lbs. The two cops got out with the body bag, and as they approached the body, they started choking. The smell was too much for them. I told them that I would reposition the helicopter to help them. I then hovered forward until the body was downwind and just below the turning rotor blade tips in front of me. I tilted the rotor back, lowered the blade RPMs so I couldn't fly, and pulled a lot of blade pitch. This created a huge and fast flow of air from the rotors, blowing the smell of the dead body away from the two cops. They stood with their backs to the rushing airflow.

With covered noses and mouths, they got the body into the body bag.

No way was the body coming back with us. I couldn't hook the body onto the cargo hook and have passengers inside the helicopter with me. That is against aviation regulations. We called Montmagny and told them to send a boat with a winch to lift the body bag into the boat.

The cops estimated that the body had been in the polluted river for five or six months.

Do you want to be disgusted? The body had several American Eels inside him when they did the autopsy. They are a favoured food along the east coast of Canada and the USA. There are eel traps all along the shallow mud flats along the south shore of the St Lawrence River.

More Trees

S-76:

The chief helicopter pilot of the Air Service came to me and wanted me to go with him to Montreal. He had met someone from the treasury department, and he told Ben that he was supposed to get $5,000,000.00 in the Air Service budget to buy a twin-engine helicopter.

Ben had looked around and found a Sikorsky S-76 in parts on the floor in a hanger near Montreal that he could get with that price. Ben wanted to see it and asked me to join him. There was just enough money in the proposed budget to buy the S-76. The S-76 could carry 13 passengers and two crew members at over 150 knots, over 170 mph.

Ben told me the budget for maintenance and operation was completely separate from the purchase price, so only the initial cost was a factor.

Good, we knew where it was, the price in its present condition, and the maintenance required to get the helicopter airworthy, all within available budgets.

Armed with as much knowledge about the S-76 as we could get, we returned to Québec City.

I needed to get my ATP and instrument rating if I were to get to fly the S-76.

I had been hearing rumours that the Congress of the USA wanted to make changes to the GI Bill. At that present time, the GI Bill paid 90% of the costs for any education at any government-approved school or university. The US Congress wanted to decrease any aviation training to 50% payment by the GI Bill because the training was so expensive that the money would be spent in just a month or two. They wanted ex-US military people to go to approved universities.

The first thing I did was send a copy of my Canadian Commercial Pilot's license to the US Federal Aviation Administration in Oklahoma City. They issued me a US FAA private pilot's license on the strength of my Canadian Commercial pilot's license. With that done, I had a record of myself with the US FAA and a pilot's license number. Getting my US ATP (Airline Transport Pilot's License) would just be an upgrade to my FAA license.

I checked around and found an approved training school called Jet Fleet in Dallas, Texas. They wanted approximately $10,000.00 US for an Airline Transport Pilot License, Helicopter, and with an Instrument rating.

I went to the Director of Operations and gave him all the information I had.

I could get over $9,000.00 US paid for by the US Government. I asked him for the other 10% of the money plus my salary and expenses to go there for the training.

He knew Ben had been looking at the S-76 and he agreed that it would be good for the Air Service to have me trained to fly it IFR (Instrument Flight Rules). If the Air Service was to buy the machine, it had to be able to fly IFR. The S-76 was certified to fly IFR, it just needed a qualified pilot to do it. Me.

A week later, the Director told me the money for the training had been approved, pending approval by the US GI Bill administrators.

I sent my request and a fax of my DD214 to Jet Fleet at Love Field in Dallas, Texas.

(DD214 is the form issued to a US military person leaving the military. It contains the person's complete military service, medals, and their honorable discharge.)

They made the application for me using my DD214, and when that was approved, they gave me a class date for the course. With that in hand, I went to the Director of Operations, and he sent a letter to Jet Fleet to have them bill the Québec Air Service for any costs not covered by the GI Bill. Everything was paid: my salary, my flight tickets, all my expenses, 90% by the GI Bill, 10% by the Québec Air Service, and all I had to do was pass the course. That included the FAA exam for the Airline Transport Pilot's License.

A month later, I went to Texas and started the course.

When I came back, I took the US license to the Director of Operation, and he had a copy put in my employee record.

Now, the Air Service just needed to buy the S-76.

All I had to do was keep my medical current for both Canada and the USA valid and up to date, then pass the Canadian IFR exam, which is very similar but harder than the US exam.

The medical was easily done. When I went for my aviation medical, it was just one more piece of paper for the doctor to sign. The Air Service paid for the medicals.

Little did I know then, but with that license, I had just guaranteed myself a job as a test pilot for Bell Helicopter in four years.

Barefoot girl:

I was called to go on a search for a young girl not far from Rimouski on the south shore of the Gaspé peninsula. I took the dog master and his German Shepard with me. The worried parents showed the police officer where they had been camping and gave him a piece of clothing the girl had worn to give the dog the scent of the young girl he was going to search for.

He spent well over two hours searching through a large open glade of trees. The dog led him to one of the little girl's shoes. Soon after he found the second shoe, the trail led down into a very bushy area along a small creek. The dog master shortly called me on this portable radio that he had seen some moose tracks. He thought the dog was following the moose's tracks, so he pulled the dog off the trail of the moose and went back up to the glade to continue deeper into the woods.

I stopped following the direction of the dog's nose and started circling around the area in ever-widening circles.

My peripheral vision spotted a flash of color, and my head snapped right, and my eyes locked on the spot of color. The color was much further along the creek than where we had been looking, and I soon saw the girl not far from the creek. She was lying down and asleep in the thick bushes along the creek. She had instinctively followed the creek.

I called the dog master and guided him to the still-sleeping young girl.

With the girl back with her very happy parents, we left for Québec City.

The dog had been following the girl, and the dog master pulled him off, mistakenly thinking the dog had started following the moose. He did not read his dog correctly. I never again fully trusted that dog master.

Lobsters:

I was called to go to the south shore of the Gaspé peninsula for a search.

It was a long day. I flew for over 10 hours, but I finally found the lost person.

The next day, the cops wanted to take some photos. When that was finished, I was supposed to leave for home. I passed by a seafood place. The cop with me was a good friend of the owner and I wanted some lobster. Several boats had dropped off their loads of lobster that morning, and the tanks were full of lobsters. I got lucky. The delivery van for the lobster place had broken down that morning and had to be towed to the garage for repairs and the seafood place was to expect many more lobsters from the boats the next morning. Driving 50 miles to rent a truck for a few deliveries just didn't make financial sense for the owner. The cop's friend wanted the room, and with his truck not available for that day and the next he gave us a really good price to his cop friend and me. I decided that since I was going straight home to take as many as I could afford.

I got a good-sized box with plastic insides, and after a layer of a few inches of ice, I bought 12 large lobsters, all in the 3 to 5-lb range.

My wife couldn't believe the price I paid and was ecstatic.

My wife immediately put her two largest pots on the stove to boil the water.

The lobster went into the bathtub where the kids were amazed and wanted to play with them. The house smelled lobster for the next couple of days.

In the eleven years that we lived in Québec City, we almost never bought any kind of seafood at the store, but our 18 cubic foot freezer usually had a large amount of all sorts of fish and seafood, what with my frequent trips to the Québec north shore and the Gaspé peninsula.

Matane, on the north shore of the Gaspé peninsula, was well known for its abundance of popcorn shrimp.

I remember along the shore of the Gulf of Mexico there were several restaurants that were close to where the shrimp boats offloaded.

Cooked shrimp were bought by bucket size and had to be shelled, which is why they were very cheap. They just had to be cooked, not processed. The large picnic tables had the condiments on them with small cups for self-made shrimp sauce. The horseradish, ketchup, mustard, tabasco sauce, and others awaited the hungry patrons. Pitchers of beer and French fries were also available. I still miss going to those places. They don't have those sorts of places in Canada.

Arrow Why:

My brother was a salesperson for a lumber company in Chibougamau, Québec. He invited my dad and I to go up there to help him with some renovations to his cottage and to do some fishing.

My parents came to Québec City. My mother stayed with my wife for a week, and my dad and I went up to Chibougamau.

I had finished the basement of my house, adding several rooms, a bathroom, a workroom, etc. I had done all the electrical wiring myself.

Dad was good with wood, and I was good with electricity.

My brother and my dad did the work on building up what my brother wanted while I crawled around under the cottage, passing all the wiring. I added a 200-amp box to bring his cottage up to house standards. He got the lumber mill chief electrician to sign it off for the Hydro Québec connection.

Then we went fishing. My brother had everything for that including a second boat that he filled with cans of gasoline and pulled behind the big boat with us in it.

We went into the park, and the game warden there had to issue the licenses.

My brother and I both spoke French, so had no problems. Our dad spoke no French.

My brother and I got our permits, OK, but the game warden could not understand my dad's accent, and when he asked him to spell his name, my dad did.

My dad's first name was Roy. In French, Roy is a family name, so he got confused and asked him again.

When back outside, we had a look at dad's license. When asked to spell his name, our dad replied, r-o-y. The warden had written Arrow Why on the license.

It took a long time for Arrow Why to live that down.

We left with our quotas of fish. My brother knew all the best places to fish.

I arrived home with another big contribution to my freezer.

WW II aircraft:

Another airplane had crashed on the north shore. The military search and rescue found the crash site from the ELT and something else they could not identify.

I was sent up to Forestville, a town halfway between Tadoussac and Baie Comeau. There, I took on an observer, and we went to the airplane crash site. We did the same as before. Got some people in there and removed the four bodies from the crashed aircraft.

We then looked for the unidentified object. It was right where Search and Rescue said it was. Turned out it was a crashed B-17 World War II bomber.

That was totally not our mission to investigate. Aerial photos were taken and a museum forensics team was sent in to the discovery to do their thing.

The Dead Book:

I worked with the photographers so often that when I finished a day's flying, I would often go up to their offices and chat with them.

One of the cops put his collection of photos of dead bodies in a book. It was about three inches thick and the most gruesome collection of photos of dead people I could ever have imagined.

Drownings, hangings, blown apart heads, car accident victims by the dozens, rotten bodies not found for awhile, murders, pieces of bodies. He had a very impressive, very morbid collection. There was one messy death that they had trouble with at first. A man had laid down in a bed beside a wall, taken a shotgun, and blown the top of his head all over the wall. The cops could not find a fired shotgun shell, so they suspected a murder staged to look like a suicide. The mystery was solved when they finally moved the body.

The impact from the shotgun pellets had caused the body to jerk up a few inches, and the autoloading shotgun had ejected the used shell. The shell had bounced off the wall and then under the bounced-up body.

One of the worst cases was one I had visited.

A murder, suicide. The house smelled of cats and death. The floors were covered with newspapers and straw. The dozens of cats had shit and pissed everywhere in the house. They couldn't get out. They had no food, so they started eating the bodies.

The dozens of cats had to be killed. The small ramshackle wood house had anything of value removed, and the house was then burned down. The smell of death and cat piss had permeated the house so badly there was no hope of salvaging the house. Only the land had any value.

Looking at the Dead Book you could almost smell the death it contained.

A few weeks later, I took the coroner for a flight when I was up in Chicoutimi. He smelled death. Every breath he exhaled smelled formaldehyde, and he was constantly sucking on breath mints. He was a nice person, but he exuded the smell of death.

I was kept busy the whole year, but most of what I did was getting repetitive, so I only added things in the book that I did differently from the routine of searches, rescues, photos, and accidents.

This was the year the Director General of the Air Service was to decide if the S76 was going to be purchased. I went to the Director of Operations and told him it would be a good idea if I got some recurrent training in a simulator. All his fixed-wing pilots, both executive and water bomber, got sim training, so why not me? I didn't need a helicopter simulator; any twin-engine fixed-wing simulator would be just fine. All I would be doing is practicing procedures. He agreed, and I was scheduled to train in the same simulator as the fixed-wing pilots.

Studying and sim time took me a week. I enjoyed the practice.

One time, I went into the helicopter hangar to talk to one of the mechanics. Those who could speak English well I always spoke to them in English. There were a couple who had trouble with English, and I spoke to them in French. One mechanic who had an English name spoke English as well as I did and preferred to speak in French. One day, he asked me why I always spoke to him in English and why I didn't speak to him in French, which he told me he preferred.

I asked him who else spoke to him in English. He told me no one.

I looked him in the eye and told him that if I didn't speak to him in English, he would soon start to lose his ability to speak in English. He paused for several seconds, told me to keep speaking to him in English, turned, and walked away.

Magdalen Islands:

The Magdalen Islands are several islands in the Gulf of St Lawrence that are 94 statute miles or 84 nautical miles or 115 kilometers from Charlottetown, Prince Edward Island, Canada's smallest province. Flying a single-engine helicopter over open water in the middle of winter scared the hell out of me.

The islands had just had their worst snowstorm in decades. Nobody could go anywhere on the islands or get to them because of the ice, and even the airport was closed for two days. It took me that long to get there.

The people at the north end of the island were completely cut off. Road closed, power and phone lines down. They only had radios for communication. We flew in essentials to them until the road and power were restored.

The main road on the east side of the main island was blocked between the town and the airport. It passed between two hills and the snow had filled the pass completely with over 30 feet of snow.

A snowblower had to cut into the mass of snow, back out and a huge payloader went in and knocked down the overhanging snow created by the snowblower tunneling into the wall of snow. It was a slow process and they had half a mile to dig out. The snowblower could only dig into the wall of snow until the column of snow ejected by the chute started to touch the overhanging snow, then back out and let the payloader in to do his thing.

Then another snowstorm hit. I had parked the helicopter behind the motel where I was staying, and because of the fierce winds, I had two 45-gallon (55 US gal) barrels of jet fuel brought to the machine, and I tied one on each side of the floats.

The next morning, the force of the winds had turned the helicopter at least 20 degrees to the right, even with the two fuel barrels and the mass of snow packed on either side.

I wanted a picture of the machine. There was a huge round rock to the right of the machine and the motel. It was over 50 feet high.

I climbed up the Lee side, and when I got to the top, the wind really hit me. I could not stand up. It was even hard to crawl into the fierce wind. For some strange reason, there was a wire fence in the bushes on top of the rock. I used that to pull myself to the windward side so I could get the picture I wanted of the snow banks around the skewed helicopter.

The mission changed the next day.

At that time, it was still legal to kill and harvest the white-coated baby Harp seals. The hunters were working from a ship west of the islands when a Greenpeace ship showed up.

The cops had to keep an eye on both of them to prevent any violence.

We were able to do that for one day, and then the worst thing happened in the worst place.

I smelled smoke in the cockpit of the helicopter. I immediately cut the battery switch and the generator and, flew to the airport, and landed as close as I could to the doors of the hangar at the airport.

I had a look in the battery compartment. A lot of fried wires. I found an airplane mechanic and had him look at it. His verdict was that the Nickel Cadmium battery had shorted internally and fried all the connecting wires.

I called the air service and told them the problem. There was a problem with my problem. They could not ship a new battery on a commercial flight with passengers. A battery was considered "dangerous goods". A special classification for air transport meant that the air service had to get me a new battery themselves. They flew in a mechanic on a commercial flight to get started on the repair. He brought many of the wires and other parts I told them were damaged.

The cops still needed the use of a helicopter, and one of the other helicopters in the air service had to bring the battery and then stay for the duration of the police requirement while my machine was repaired. A lot of wiring had to be replaced along with the battery. When I got the machine back to the Air Service hanger in Québec City, it took another few days to complete the repair. Repainting the scorched metal would wait for a major inspection to complete, so they did only the minor touch-up on the outside paint so the damage didn't show.

The cops had for years given me an unmarked patrol car in the evening so they didn't have to taxi me around to motels, restaurants, or wherever else I wanted to go. They didn't have one to give me on Magdalen Islands, so they gave me a marked car with everything on it.

All told I was on the Magdalen Islands for over a week.

Ice flow:

I brought the helicopter back to Québec City with the scorched metal in the nose. The blackened wire going up from the battery through the windshield separator to the relay in the engine compartment, which stopped the power surge from getting to the starter generator, also had to be replaced. In addition, a 1,000-hour inspection was coming due, so the maintenance supervisor grounded the helicopter, and I had to use the machine normally used by the Premier of the province. His machine had police radios in it and was often used by many of the ministries as a replacement aircraft. My machine was down for maintenance, so I used that one for a couple of weeks.

I got a call to go get three young boys on the river.

They were a mile or so west of the Twin Québec bridges. I saw them right away as I came to the river. A passing motorist along the north shore highway had spotted the boys that were playing among the ice pieces brought in by the last tide. Most of the pieces of ice left were about 10 to 20 feet wide and 2 to 5 feet thick. The boys had been about 50 yards offshore when the tide started coming in. They all got on an ice flow, and the waters soon surrounded them, making it impossible for them to get back to shore. Without rescue, they would have floated right past Québec City.

I approached the ice flow the 3 were on, and about 50 feet back, I came to a hover. Reaching behind me, I stretched back and opened the right rear passenger door. I was tall enough and the seat back low enough that I was able to do that. I had done so many times in the past. It helps to be tall.

The police helicopter had large fixed floats that would have been perfect for this rescue, but the machine I was flying had low skids mounted under it.

I hovered up to the edge of the ice flow with the door open, and the three boys stepped into the back seat. As each one entered, I had to adjust power to compensate for their weight as each one stepped into my hovering helicopter. When all 3 kids were aboard, I reached back & closed the right rear door.

There was a police car waiting along the highway, and I flew to him. The three boys were helped out by the cop and taken to their parents.

I landed back at the air service and was cooling down the engine prior to shutdown.

The Allison 250-C20B engine required a two-minute cooldown at idle speed to bring the engine oil and bearing temperatures down to stabilize the engine at a cooler temperature prior to shutdown.

While I was waiting the two minutes I heard an Air Canada pilot request clearance to land. The tower told him that a small Cessna was ahead and below him on his landing approach to the runway. The Air Canada pilot then asked if the tower could tell the Cessna to make a go-around so he could land first. The tower said no. The Air Canada pilot then asked him a second time, saying that his expensive passenger airliner cost $100.00 a minute to fly.

The control tower operator came back without missing a beat and said, "Air Canada 123, make an immediate $200.00 turn to the right and call when established on final".

The Cessna landed several times. The Cessna must have bounced 3 or 4 times. I gave the Cessna pilot the benefit of the doubt. He bounced because he was laughing so hard as he was landing, and not from inexperience.

Suicides:

I did a lot of flying that spring, and it was a year of flooding. Lots of stranded people, a few drownings, and one suicide that was different. A man threw himself off a bridge, and downstream, he was lodged in some rapids. The water was moving too fast and was rocky and deep enough to prevent someone from going down in a boat or wading into the river during flood season.

I had to land on the water very lightly so the force of the flow would not carry me downstream.

The police officer had to climb out on the float with a rope that had one end hooked to the cargo hook and the rest of the rope in his hands inside. He had to try and try to lasso one of the legs of the victim. He did so after three or four tries.

He got back inside, and I lifted the body clear of the water and hovered over to the shore and the waiting ambulance. The body hanging upside down from the cargo hook of the helicopter by one leg must have been a grotesque sight.

Two weeks later, I was again called to the south to a small, deep quarry. A man had jumped from the edge of the quarry into the water over 100 feet below.

His body came to the surface and was spotted by a searcher.

A cop had to climb down the steep road to a rock pile near the shore of the water and wrap a rope around the shoulders of the victim. After that, the cop hooked the body to the cargo hook of my helicopter as I hovered over him, being very careful not to hit the steep rocky side of the quarry with my rotor blades. I lifted the body out of the quarry and up to the top on the side where an ambulance was waiting. The only clear landing place I had near the ambulance was a sandy area. As I hovered towards the ambulance, all the spectators pulled back from the edge of the quarry to give me the place to land.

Except one person.

He turned his back, bent over, and covered his face against the blowing sand. I hovered right up to the idiot and he didn't move. The sand was blasting him, and he was holding his shirt over his mouth with his eyes screwed shut.

I waited a few seconds, then lowering the body to the ground, I hit the idiot in the back with the nose of the float. He stumbled forward a step and again just stood there, covering his head and not moving. I backed a bit, then pushed forward and hit him again, harder. This time, I knocked him over onto his hands and knees. He finally got the message and scrambled away from the sandy area onto some grass. I brought the body close to the ambulance and let the body settle to the ground again, where I then dropped the rope from the hook.

Hughes 500:

A Hughes 500 is a type of helicopter. A smaller and faster turbine engine machine with 4 blades and carried the same number of people as my Bell 206B but with less room.

That summer a Hughes 500 crashed up near Indicator Lake, north of Chibougamau. There were five dead, and all were underwater. I went up there with two scuba divers.

It was a sad case. The pilot was showing off late in the afternoon and had flown into a mirror. Winds were calm, and it was just before sunset. The pilot had the sun in his eyes. The pilot of the Hughes 500 had buzzed a camp with several girls in it and then popped up and over the hill behind the camp. He then dived back down on the other side of the hill into a bay on the other side. He flew into the flat, calm waters with the sun in his eyes at over 120 knots. Bad luck and poor judgement.

The divers got the passengers out of the sunken machine and tied a rope to the rotor hub. With this rope, a more powerful machine was able to pull the Hughes 500 to the surface of the water and slowly drag it to shore, with water pouring out from the smashed windshield and twisted doors of the machine.

The five bodies and the diving gear belonging to the 2 cops were then loaded onto an Otter bush airplane on floats, and the pilot then flew them down to Chibougamau.

I had a chance to talk to the chief pilot of the company that owned the Hughes 500 and the more powerful Bell 206L that had pulled the 500 to shore.

He told me he had a contract in James Bay at the La Grande 2 construction camp, and the dead pilot was supposed to do the job.

I told him that I could take some vacation time and do three weeks of the contract.

He gave me the information I needed, and when I got back to Québec City, I asked for one week's vacation. Add that to the week before and the week after my vacation week, and I was able to work for him and his company for three weeks.

I finished my week of work on Monday evening, and on Tuesday morning, I drove to Montreal to get a Pilot Proficiency Check from their instructor pilot. Then, they put me on a plane to LG2 in the James Bay complex. The contract was to work for hydrology. They were testing water flows and temperature at various depths in the LG2 reservoir.

I even got some time to do some fishing, and when the customer wanted to go fishing along a major rapid, I stayed close to the helicopter. The other three took off further along the river. By the time they came back, I had caught four speckled trout of between four to five pounds. They caught nothing.

I was doing my hydrology job but one of the crew was a young lazy bastard and would unload light stuff from the helicopter and put it beside the machine. This was dangerous because the rotor wash could pick up light objects and then lift them into the rotor blades. After having to tell him a few times to move the light stuff at least fifty feet from the helicopter, he pissed me off. The aviation safety officer for the LG2 project was available and I told him my problem. I told him also that the lazy kid was unsafe around a helicopter, and if he didn't do what I told him to do I would refuse to permit him aboard any machine that I was flying. The safety officer talked to the kid. If I refused to let him aboard my helicopter because I thought his actions were unsafe, he would be fired and put on the next plane to Montreal. Very reluctantly, he obeyed me after that. The kid always made a show of extreme effort every time I ordered him to move something.

A few days later, we went about 60 miles south with the crew for a hydrology job. It was a hot, sunny day, and I felt vindictive. The kid had taken off his T-shirt and left it in the cargo compartment, unreachable in flight. We had a long way to go to get back to the LG2 camp, and it was hot. I climbed to 6,000 feet. At three degrees per thousand feet, it was a lot cooler up high. The kid was shivering. Good.

When I got home, I went to the photography guys at the police headquarters and asked them what would be the best camera for taking aerial photos. They suggested a model of Nikon with a 43-86 mm lens. That camera was with me all the time in the cockpit after that. The Nikon eventually paid for itself at least three or four times.

I also found a way to get free film.

The cops always used black and white film because they could develop the film themselves. I always kept black and white film loaded in the camera. When pictures were needed, I took them and then traded the roll of pictures they wanted for two rolls of black and white and one roll of color film. I got what I wanted, and I saved the police a lot of time and money sending out a photographer.

(Later, Denis Lortie and the Pope were really the cherry on the payment of the camera.)

Sunken outboard

There was a confrontation with an Indian tribe in the Gaspé peninsula. They were complaining to the government about something and had blocked the bridge crossing the Matapedia River from Québec to New Brunswick at Campbelltown, New Brunswick.

The Indian tribe police chief had tried to get the Indian militants to stop blocking the bridge, and the militant Indians retaliated by stealing the Indian police chief's new 50 hp Johnson outboard motor worth several thousand dollars.

In the middle of the night, using bolt cutters, the Indian militant thieves cut the locks and removed the outboard motor from the police chief's boat. The cops canvassed the area, looking for information. A woman had witnessed the militants taking the motor out into the middle of the bay and dumping it into the water.

The chief's futile search over the water from a boat for his new motor finally led him to the Provincial Police.

I had been there for a day already, and the police asked me if I could help the Indian chief find his outboard motor. Sure, why not?

The Indian arrived with a cement block and a long piece of rope tied to an empty plastic gallon jug.

It took me about 2 minutes to find the best way to search. At 100 feet and about 20 knots, the bottom of the bay was quite clear. It took me about 20 minutes in the large bay to find the motor in about 50 feet of clear water. I landed on the water over the motor and the chief dropped the cement block down beside the motor.

The chief quickly got a diver to go down and tie a rope to the motor, which he was able to recover.

I finished flying for the day, and the police dropped me off at my motel after dinner. That night, in my motel room at about 9 pm, I heard a knock at my motel door. The Indian chief was there with a good-sized box.

He thanked me again for finding his motor and handed me the box.

After he left, I opened the box, and laying on a thick bed of crushed ice was a fresh, cleaned, 20 lb salmon.

What a way to keep my wife happy. She loved salmon steaks.

CFFPP:

I got called for a search in Chibougamau.

After refueling in Roberval in the Lake St Jean area, I continued on to Chibougamau. Halfway there, I spotted an aircraft in pieces on the side of a large swamp.

The wings and tail were removed from the fuselage, and the aircraft looked otherwise undamaged. It had been in the swamp for a while. I landed beside the cockpit and saw that all the instruments and the engine had been removed. I noted the aircraft registration painted on the wing, CFFPP, and then continued on to Chibougamau.

The next morning, I found the lost person in about an hour, refueled and went back to Québec City.

The next day, I called the Minister of Transport Office (MOT) in Ottawa and asked a few questions, and gave them the registration.

MOT called me back and he told me they found an aircraft with that registration in a hangar in Hull, Québec. (Across the river from Canada's capital city, Ottawa)

That was unusual. Normally a crashed aircraft had their registration letters retired, never to be used again. That was MOT's problem now.

Curious, but then registration letters are expensive to get.

Five and three:

The helicopter was needed north of Roberval. Five fishermen had wandered into the bush and not returned. The cops took me to the truck the fishermen had arrived in. It was parked beside their camp.

It was a clear, warm day, and I circled the area for a couple of hours and at one point I hit the siren to let them know that the police were looking for them.

That night, a front moved past and left behind rain, low clouds, and thunderstorms.

I returned to the search area, and after about an hour, I saw a column of smoke reaching for the low clouds. I found them along a creek, and they were all on top of a beaver dam.

I had burned enough fuel to be able to lower the helicopter vertically to the top of the beaver dam and pulled the two biggest ones out. I dropped them back at their camp, where a police car was waiting for them. I went back and got the other three. I asked them why they didn't build the fire the day before when it was clear and dry. One guy said they heard the siren and thought we saw them.

At least they got out alive.

The next three were not so lucky.

Near Chicoutimi I found their boat, half full of water, floating on a small lake. We went back to Chicoutimi and picked up a diver.

The diver found the three of them under the boat. On a metal fish chain, he also found several fish, still alive, wired to the boat.

They had a large rock as an anchor tied to the oarlock on the side of the boat.

The rock was too heavy; the rope was too short, and tied to the oarlock in the center of the boat tipped them over when they dropped the rock overboard. They were also not wearing any life vests or any other floatation gear.

Remove any one of the four mistakes they made, and they would still be alive.

Nuclear plant:

The only nuclear plant in Québec was on the south shore of the St Lawrence River, just east of the city of Three Rivers. The administration of the Gentilly Nuclear plant wanted to stage a practice emergency with the Provincial Police. The nuclear plant was free of any habitation within a couple of miles.

They all went through several emergency scenarios and a couple of them required the helicopter to assist in their simulation. I got a tour of the place for my efforts. I was given a little badge that counted the rads that I was subjected to during the tour.

Practically nothing, so I didn't have to worry about having two-headed babies.

On the way back to Québec City, I ran into a line of thunderstorms just before I got to the airport. It was getting dark, and the thunderstorms made it darker. I followed the highway at about 200 feet as I entered the rainstorm. A bolt of lightning flashed just ahead of me. Too damn close, it scared me. By too damn close, it was less than 200 feet ahead of me. I couldn't see more than a couple of hundred yards in the heavy rain when the lightning struck. The only thing I could see was a bit of the south shore. I immediately turned south, crossed the river, and followed the shore to the twin bridges. There, I was just ahead of the thunderstorm, and I turned north to my base at the Air Service.

I got there, did my cooldown, tied down the blades, and got into the hangar door just as the rain hit.

I have heard of aircraft being struck by lightning, and I did not want to experience that happening to me.

I later found out that all aircraft are designed like a Faraday cage to turn the lightning along the outside skin when it is in the air.

North shore:

Rene Levesque had been elected the premier of the Province of Québec.

He led the Parti Québecois, which had to dedicate itself to separating Québec from the rest of Canada. (After two referendums, it didn't happen.)

Rene wanted to visit the Québec north shore.

The distance from Québec City to Blanc Sablon is over 900 miles or 1,400 kilometers. The road ended at Natashquan, over 300 miles from Blanc Sablon, which is at the far eastern tip of Québec, just across the strait from Newfoundland.

The land was grey granite. No dirt and almost no trees.

We left Québec City with Rene, his wife Corinne, and a bodyguard.

We were gone a week what with all the political stops we had to make. Two other bodyguards went ahead via airplane to arrange accommodations and protection.

When we got to Blanc Sablon, the small detachment of police officers invited us all for a snack of smoked salmon. They had a salmon of about 25 pounds all ready. It was delicious. I sat on the sofa between Rene and his wife Corinne with six police officers. I got a kidding because I was English, and they all managed to pick on me in good-natured fun.

After the delicious snack, the cops showed us the smokehouse. They had a small hut with a wood stove and several cords of imported maple wood. (The only trees they had within many miles were pine trees. Pine burns like paper.)

The salmon were in another hut about 100 feet away, with the chimney running between the two. The smoke was cold by the time it got to the fish, and they smoked the fish for a couple of days.

One thing about Rene Levesque was that as much as he was separatist he was always very polite with me and always spoke English with me when we were alone or on the helicopter intercom.

The Nun:

The emergency squad guys wanted to practice some rappelling. I went with them but the instructor was not too willing to have me jump. I told him that I had done rappelling in the US Army, so he let me jump off the tower a few times. He saw that I knew how to do it and was satisfied. We also practiced other life-saving techniques including a fireman's carry. They also had a cradle stretcher that could be lifted under the helicopter. The police loaded it with sandbags, and I took a circuit around the headquarters building. I lifted it with a 50-foot-long line attached across the back floor, not the cargo hook, so it could not be dropped.

A couple of days later, I got to practice some of what I learned for real.

A nun had been going for a walk on the south shore along the edge of the cliff under Québec's twin bridges. The ground had given way under her feet, and she slid down the nearly vertical cliff.

A passing ship spotted her lying at the base of the cliff. We were sent to get her. I landed beside her, and she told us that she had broken her ankle. I was with one of the emergency squad members that we had practiced the lifesaving procedures with a few days before. She was a big woman and in her 60s at least.

My observer and I had little time to waste; the tide was coming back in, and I knew the tide was between 10 and 12 feet near the bridge.

I had learned that when I rescued the three boys on the ice flow.

I opened the left passenger stretcher door, and my observer and I picked her up using the fireman's carry we had practiced just the week before.

She was a tough old girl but the pain in her ankle had her crying.

We got her strapped in, and she said she was scared; she had never flown before. That's OK, I did the same thing I did with the bishop. Upon liftoff I had her look up at the bridges and then the airport, the old fort, anything on the horizon. I didn't give her time to get scared by looking down.

We got to the hospital, and they were ready for us. My observer had followed my instructions and had them send only one person to the helipad with a wheelchair. I did not want a replay of the dangerous scramble that had happened with the incubator baby.

My observer and I got her out using the fireman's carry. Again, with our hands locked to our wrists under her knees and across her back.

She had her arms around the necks of my observer and myself. We got her into the wheelchair, and since my head was locked in a headlock, she pulled my head towards her and gave me the biggest, wettest, hardest kiss on the cheek that I had ever had in my life. Talk about getting a holy kiss. The nun had probably never kissed a man since she took her religious vows so this was her one and only opportunity to do so in her life, so she planted the kiss with a lot of gusto. The cop, seeing this, unwrapped her arm from around his neck before she could do the same to him. She turned and waved a hearty goodbye as she was being wheeled into the hospital. She had just experienced one of the most momentous events that had ever happened to her in her cloistered life.

Two weeks later, I got called to the director's office and he showed me the thank you letter from the mother superior of the convent the nun was from.

The director later tacked the letter to the bulletin board of the pilot's waiting room. I got a few ribald comments for that from some of the other Air Service members. One of them was that I couldn't find a date, so I was chasing nuns.

Compass error Manic 5.

A couple of days later, I got a call to go to Baie Comeau. It was late in the afternoon so I was unable to start a search for a boat with two men in it that were missing for the past four days.

The next morning, I went up to the Manic 5 reservoir. It is one of the largest freshwater reservoirs in the world, over 1,150 feet deep and holding 34 cubic miles of water with a surface area of 750 sq mi, or 1,942 sq km. The Three Gorges Dam in China was the only larger man-made reservoir in the world.

Manic 5 was formed over 212 million years ago by a very large meteor. The island in the middle was formed by the molten rock bouncing up after the impact, much like a stone would have the water lifting into a cone after the impact. It is one of the largest known terrestrial impact craters in the world at 40 miles, 65 kilometers in diameter.

It is over 60 miles (95 kilometers) from the dam in the south to the north side of the reservoir.

The boat I was looking for was a good-sized cabin cruiser.

The island in the middle of the Manic 5 reservoir is over 1100 feet high, so I climbed to several thousand feet to be able to see both sides of the donut-shaped reservoir. At the very far north end of the reservoir, I spotted the boat.

I circled over their boat and landed on a wide sandy beach close to them. The owner of the boat and his employee came in to meet us.

He had almost run out of gas. They had just enough gas left to power a small lifeboat with a 5 HP engine. He had tried to find the entrance to the Manicouagan River to the Manic 5 dam. I told him it was 60 miles south. He told me he was at the south end.

He was looking for the river entrance in the fog and had passed it. He was convinced that the compass had turned 180 degrees in the fog.

I showed him where he was on the map. He didn't believe me. He was convinced that he was at the south end.

He didn't believe me, so I asked him where the sun was; he pointed south and called it north.

His mind was really bent. I told him you are lost, not me.

I got them both into the helicopter and flew back at several thousand feet so he could get his mind straightened out.

He had to rent a float plane to fly them and fuel back to his boat.

I told him when we parted to always believe his compass. The only thing wrong with his compass was in his head.

The magnetic compass does reverse itself, but it hasn't done so in the last 780,000 years.

The next week, one of the CL215 pilots brought his machine back for maintenance.

He later had to take it for a test flight to do his systems checks. He invited me along.

He got me into the co-pilot's seat, and I spent over 45 minutes flying that big beast. He let me try everything except a water pickup, which was interesting. I found that in a roll, it was slow and heavy to control, whereas nose up and down was much easier and responsive. He took the controls and did a water pickup and drop.

The CL215 can pick up and drop 1,200 imperial gallons (12,000 lbs) or 1,440 US gal of water. (5,455 liters). The CL215 successor is the turbine-powered CL415, which can carry 1,350 imp gal (13,500 lbs), 1620 US gal, or 6,140 l.

One of the most unique jobs I ever heard of them doing was in Chicoutimi, in the Lac St-Jean area.

A large shopping center had caught fire between two big box stores anchoring the ends of the shopping complex.

The manager of one of them, Sears, did not want to lose his store.

He called in two big bulldozers and the CL215's.

The bulldozers cut the Sears store off from the burning part of the shopping complex. The water the firemen were putting on the building never got to the fire. The roof protected it. The bulldozers solved that because the building was too wide for the firefighter's water hoses water to reach the center. The water bombers fixed that and the fire was contained fairly quickly, then permitting the firemen to get their water hoses to the center. With the fire out, workers went in and blocked the entrance to Sears from the burnt and smashed center of the shopping complex.

In 2 days, Sears was back open for business.

I did not see the water bombing, but I did see what was left of the shopping center a few days later. Most of the building was a scorched and blackened mess, but the Sears store was open for business.

Tadoussac:

I was called to go to Tadoussac, where the ferry boat crews were on strike.

The mouth of the Saguenay at Tadoussac has two ferry boats that cross the river on a varying schedule depending on traffic, often every 30 minutes. They are, in effect ,part of the highway, and it's free to cross. The mouth of the river is 1 to 2 miles (1.5 – 3 km) wide, the water is over 1,000 feet deep at the mouth of the Saguenay, and the hills go right up to 1,500 feet immediately upstream, negating the possibility of an affordable bridge. From the top of the hill to the water was almost sheer and then dropped sheer again another 1,000 feet. Anything going to the Québec north shore had to fit on the ferry or it went by boat.

The day I went up there, it was very windy, which is common in that area.

I was following the shoreline northeast bound at about 1,000 feet and the strong winds pouring over the hills were very turbulent. I elected to decrease altitude and get low along the railroad tracks carved into the shoreline of the river. I couldn't go down. The wave of wind was pushing me up. I rolled the engine to idle and entered autorotation, but was still going up. I had to veer left and hug the trees along the side of the hills. That let the wave of the wind pass over me, and I was able to go down. Rolling power back on I stayed about 50 feet above the river all the way to Tadoussac.

I started ferrying police and other dignitaries across the river. The Coast Guard base was also blocked by the strikers, so the police brought fuel via truck to an open parking area a couple of miles east of the river. There I refueled, and three large police officers wanted to go down near the ferry. The hill down to the ferry was very steep, very long, and blocked by the strikers. Not even the cops could get past, which was why I was there.

I was taking them to a small flat, open area down near the ferry. The area was on the edge of the hill, and the winds were flowing downhill, so when I went in for my approach, I was committed.

I checked my vertical speed indicator, and it showed me climbing when, in fact, I was maintaining altitude approaching the landing area. Now I was worried. I was maintaining over 90% power just to hold my altitude and I was still in the translational lift. It was obvious that if I came to a hover, I would over torque the transmission. For the last 20 feet, I pulled 100% power and let the helicopter settle toward the ground. A hard landing was coming up and I would damage the machine if I let that happen.

Descending quickly at 100% power, I chopped the engine power at about 5 feet, arresting the descent, and did a hovering autorotation. It was noisy and rough, but it stopped the impact. It was a safe, noisy, abrupt, and very unconventional landing. I was scared and shaking. What I had just done was not in any manual, and if I had realized how strong the descending winds were, I would never have attempted the landing. Poor prior planning on my part.

The three men got out, and now I would have enough power to take off minus the almost 600 pounds of meat and bone.

That night, the strikers and the government came to an agreement, and traffic started flowing again.

As I said at the beginning of the book, experience is a hard teacher; it gives the exam first and the lesson afterwards.

Hells Angles:

A Hells Angels president died, and the funeral was huge. Hundreds of full patch Hell Angels members came to attend the funeral in Drummondville, Québec. (Southeast of Montreal.)

When the Hells Angels ride their motorcycles in groups, normally with a Harley Davidson motorcycle, but any US-made motorcycle is permitted. They always wear their colors which is a vest with the Hells Angels patch on the back when riding together. There were many hundreds of bikes at the funeral.

The cops frequently stop them to check and search for drugs and/or weapons. The smart ones have their women bring the drugs and weapons in a car or van and connect with them upon arrival at their destination. Some would choose to ride the back roads, and that's where the helicopter came in handy.

They would wear a patch with 81 on it. H was the 8th letter of the alphabet, and A was the first letter for Hells Angels. The American Motorcycle Association once let it be known that 99% of bikers were honest, legitimate motorcycle owners, so the Hells Angels prided themselves on being the 1 %ers.

They caused no trouble but took over every hotel, motel, campground, and restaurant for several miles around.

Kid in park:

I had been on a lot of searches over the years, and I had been able to train myself to use my peripheral vision.

Many times, my head would snap around to have me looking at the person or people I was searching for. That was happening on a regular basis as I trained myself to see with my eyes what my mind was seeing.

That really increased my success rate on searches, and with the bloodhound's nose, my success rate on searches was getting close to 100%.

I was called north to Jacques Cartier Park, where a 14-year-old boy was lost.

I had no dog this time and spent about an hour circling and following the valleys and all the streams I could find in the area where the boy's family was camping.

I turned towards another creek, and suddenly, my head snapped up, and my eyes locked on a piece of tangled bushes along the watercourse. I did not see him, but my eyes were locked onto a spot. I kept my eyes locked on the spot and flew directly towards it. It was almost two miles away. I couldn't believe it. I saw nothing with my eyes. Many seconds later, as I bored into the location my eyes were locked onto, I spotted movement. The boy was along the stream in bush, almost up to his shoulders. I was amazed that I had seen him from that far.

My flying 6th sense had really kicked in and spotted the boy.

There was no place to land, but I hovered down as low as I could get, putting my floats into the bushes, and the boy climbed up onto the float, opened the door, and got in. With the door closed and the boy's seat belt attached, I flew to where a cop and the boy's parent were waiting.

For me, it was an incredible, mystifying experience. My mind, my 6th sense, found the boy.

When I got back to base, I met Ben, the chief pilot, in the helicopter hangar.

He looked glum and told me the bad news. He had just been talking to the Director, and he was told that the $5,000,000.00 to buy a twin-engine helicopter had expired in the last budget. The Air Service was going to contract all helicopters except for the police. They were owned by the Ministry of Justice and the Air Service just supplied the pilots, mechanics, and the hanger.

Any twin-engine helicopter future for the Québec Air Service was dead. The one positive thing I had out of that dream was my US FAA ATPLH IFR rating

(United States Federal Aviation Administration, Airline Transport Pilot, Helicopter rating with Instrument Flight Rules rating).

I had been taking a lot of pictures since I got my new camera. I had gone to the two newspapers in Québec City, and one was just not interested in aerial photos.

The other newspaper was mildly interested, and I only sold them three or four pictures at $35.00 each. They paid me cash, and I gave them a receipt.

That at least got the ball rolling.

I had spent a lot of time with the police photographers and had learned how to develop black and white films. They had their own lab, so it was easy. They had all the equipment and photo paper, and I learned how to use everything. Several times during my seven days off I would go to the police HQ and hang out with different groups, learning their operations and how they worked. They got to know me quite well and taught me a lot, so when I worked with them, they had complete confidence in me.

NPBA

The police have many different departments, and I learned most of them. I spent a lot of my free time at the cop shop with the various cops with specialties. Snipers, I shot with them. The shooting instructor, I shot every type and model of pistol and submachinegun in their firing range, and he helped train me. Divers, I learned their needs and equipment weights and worked with them often. Photographers, I spent a lot of time with them and their equipment. Dog master. The one with the bloodhound became my friend, and I went with him when he worked with his dog.

He had become a member of the NPBA, National Police Bloodhound Association in the US. He told me he wanted to go to an NPBA convention in Cincinnati. He couldn't speak English so he asked me to go with him. We went and got permission for both of us from the Emergency squad Director and my Air Service Director.

We got a lot of very odd looks from many who saw a marked Québec Provincial Police car on a US interstate highway on our way to Cincinnati, Ohio. The dog master didn't bring his gun or his dog.

The various bloodhound dog masters had a lot of information that they shared with us. During one big meeting, they wanted him to go up on stage to answer some questions many had for him. He couldn't go alone; he couldn't speak English, but one curious thing was that all the dog's commands were in English.

I had to go up on stage with him while they asked their questions in English, Texan, and Brooklynese. I told him the questions in French and repeated his answers in English. It was an amazing experience. It was my first time on stage and my first time as a translator on stage. When I went up on the stage, I was very nervous, but that soon went away, what with my need to concentrate on translating the two languages. Dog handling is a science, and it was a great learning experience.

Rimouski boat.

I was called to go to Rimouski again for another search and some aerial photos.

It was a windy and turbulent day, so I kept low over the waters of the St Lawrence River. The only reason I could do that was because of the large bags of the fixed floats mounted under the helicopter. They slowed the machine down and reduced my maximum gross weight, but because I had to fly over water so often that they were essential for a single-engine helicopter.

With the job done, I landed at the airport to refuel and then went to the police headquarters in Rimouski. On the way there, I saw a newly installed sign at the newly built Hydro Québec building right beside the police HQ. The Hydro Québec logo is a large Q with the line on the bottom right of the Q showing a bolt of electricity. On the sign, the electricity bolt was on the left side. The worker who had put the sign up had installed it backwards.

It was a stupid mistake. Publicly very embarrassing. I took a picture of it, and when I got to the police headquarters, I went straight to the Director's office and showed him the photo. I asked him to come with me and we went to the front entrance and from there he could see the sign for himself. He never cracked a smile, but he did call the Hydro Québec director right then and told him what a stupid, incompetent job had been done with his sign. The HQ director was instantly aware of the public relations debacle that could cause.

It was getting late in the day, and it was very windy. The winds were from the northwest, and when they hit the land, the wind boiled over like water in a rapid. I decided to fly low over the water on my way back to Québec City. The wind speed was faster but with no turbulence. The air passing over the water had flattened out due to the lack of resistance by obstacles. The winds were very strong, over 40 miles per hour, but with only 10 to 20 degrees of headwind component coming from the right with very little turbulence.

Several miles later, along the north shore, there is a long sheer cliff dropping straight into the water from several hundred feet high. About half a mile from the cliff a single-masted sailboat was anchored and heaving heavily into the blasting winds and foaming waves.

I circled the boat and waived to the people on board. I could do nothing to rescue them. The boat's mast prevented me from approaching the boat. I got on the police radio and called the cops. They could do nothing but send a patrol car to where they could keep an eye on the boat, from over 2 miles away. I came to a low hover beside the boat so they could see POLICE painted on the side. At the hover, my airspeed indicator was bouncing just over 40 knots. (40 X 1.15 = 46 mph or 40 X 1.852 = 74 kph)

I again got on the police radio and requested a conference call to the dispatcher at my base in Québec City. I told him that this needed the Coast Guard and gave him the coordinates of the sailboat. They could do something.

I was over water trying to give an accurate location of the sailboat using an 8-mile-inch aeronautical map with one finger while flying with the other hand. A 1/8th of an inch was a mile and over water. Overwater distance, guesstimating was extremely difficult, but that's all I could do; I had no reference point to guess the distance.

I couldn't stay over the boat. First, I didn't have the fuel for a long vigil, and it would be dark by the time I got to Québec City if I stayed any longer.

Over an hour later, when I was approaching Québec City from the east, I saw a Coast Guard ship looking to be moving east as fast as he could.

The next day, I found out that the sailboat had been rescued and shielded from the winds by the Coast Guard ship, and they had been towed to safety to the Rimouski harbour.

Half the crew was on shore leave for the day in Québec City, but the Captain didn't wait a minute when he got the call. He untied from shore and went full speed for the rescue.

The Coast Guard cutter arrived well after midnight, and with his ship lit up like a Christmas tree, the Captain located and rescued the occupants of the sailboat and their boat.

It would have been amusing to see the bewildered faces of the crewmen when they returned from shore leave and saw the empty berth along the Coast Guard docks in Québec harbour.

Deer on ice.

In mid-January, we were out taking some photos, and we spotted something on an ice flow between the Québec harbour and Orleans Island to the east.

A deer was lying on the ice in the middle of the river. As we approached, the deer was nervous and agitated but did not get up. She was terrified of the helicopter but she was also terrified of leaving her ice flow in the middle of the ice-choked St Lawrence River. I flew the helicopter to the shore, and my observer got out and went back to the cargo department of the Bell 206B and came in with a rope. I made a loop at one end for the cargo hook and a lasso at the other end. I told him to hook the rope onto the cargo hook and then bring the rest of the rope with him into the cockpit.

We hovered back out to the fast-moving ice flow in the middle of the river. I lightly put my float on the ice flow beside the deer. She was shaking with fear, and her eyes were huge in her fright, but she didn't move. My observer put the rope around her shaking neck and managed to get one leg into the loop of the lasso.

He pulled the rope tight around her leg and neck. Looping only her neck would have caused the rope to strangle her when I lifted her off the ice flow. As I started to lift her from the ice flow she was by now extremely agitated and got up and tried to fight off the rope, but to no avail. I lifted her from the ice and she was still fighting the rope. She stopped fighting the rope as I hovered over the ice-filled water towards shore. Watching her in the mirror I gently lowered her to the ground. I gave the rope lots of slack and with her feet firmly on the ground, she was able to shake herself free. Unhurt and on land, she bounded off towards the trees. We landed and the cop got out and put the rope back in the cargo compartment. The cop and I traded looks and smiles. We were both very happy to have saved the life of the female deer.

During my life, I have shot several deer but the satisfaction of saving the life of that deer was far greater than my shooting a deer ever was.

Denis Lortie

Denis Lortie was a corporal in the Canadian Armed Forces. He was based in Carp, Ontario, west of the Canada's capital, Ottawa, Ontario.

He was a soldier trained in the use of weapons and first aid. Denis Lortie was a military supply technician, meaning he had access to any portable weapon he wanted.

On May 8th, 1984, twenty-five-year-old Denis Lortie arrived at the Québec City parliament building with two C-1 (L2A3) Sterling submachine guns in 9mm plus an Inglis 9x19mm pistol with a 13-round magazine (Browning Hi-Power) and a knife. He also had a small bag with a lot of 9mm military-grade ammunition, a first aid kit, and a gas mask.

Denis Lortie entered the Québec parliament building on the south side, parking his car in front of a southside service entrance. He went into a small anteroom and murdered three parliamentary pages with the machine gun. He then opened the door, entered the parliamentary chamber, and shouted "Where are the MNA's, I want to kill them."

He opened fire with his submachine gun spraying the Parti Québecois side with dozens of rounds of 9mm bullets.

In total he killed three people and wounded thirteen others.

He had entered the door facing the chair of the Speaker of the House. The government MNA's on the right of the speaker were the members of the Parti Québecois. On the Speaker's left are the opposition parties.

Denis Lortie continued shooting dozens of bullets into the Parti Québecois side, shooting into the seats with a steady stream of fire going through several clips of ammunition until he reached the Speaker's chair.

The parliamentary chamber was almost empty. The 09:30 start of parliament had been delayed by 30 minutes. He had entered at 09:45. He had failed.

Denis Lortie sat in the Speaker's chair. He looked back and saw the camera facing him over the door he had entered. Denis lifted the Sterling submachine gun and sprayed the camera with 9mm bullets. He missed.

I was in the photo department at the police headquarters building.

The photographer on duty got the call. Murders at the parliament building. People were dead, killed with a machine gun. Grabbing his photo gear, he yelled for me to follow him.

255

We ran out to the helicopter parked beside the police headquarters. After start-up, we took off and I called the Québec Airport Control Tower and declared a police emergency exclusion zone of 3,000 feet above and 3 miles in diameter to give me an exclusion dome around the Québec parliament buildings.

This did two things: it would prevent any aircraft from approaching and possibly getting shot at, plus it gave me my own control zone within a control zone where I would not have to worry about any other aircraft from approaching and getting in my way.

This was almost like when I was with the cop who shot the moose inside the airport boundary, only this time the controller retained control of his airport and only relinquished a small part of his control zone to me.

The photographer kept taking photos occasionally to record any changes as they happened.

I had been over the area for over an hour when a charter aircraft arrived overhead. The pilot knew my name and asked if he could drop down for some pictures. I had to refuse him.

About half an hour later, we got a call that Lortie had surrendered to the police. The Sergeant-at-arms of the Parliamentary Chamber, Rene Mark Jalbert, had talked Denis Lortie into surrendering without anyone else getting injured.

We went back to the cop shop and landed.

The photographer loaded up with several more rolls of film and we headed back to the parliament building in his car.

We went in the main entrance and through the anteroom where 3 people had been murdered. The walls had several bullet holes in them and a large glass-topped table was awash with blood. I have seen many gruesome sights but the fresh blood on the glass table was a sickening sight to behold. The deranged mind of someone who was capable of murdering three young, unarmed people was a taste of what was to come. We went through the door to the main chamber and were stopped and told to go up the stairs to the visitor's gallery. Only the police photographers were allowed on the main floor which was speckled with dozens of 9mm bullet shells from Lortie's Sterling machine gun. I stopped and looked for a very long moment. I would never see something like this again in my life.

One thing I did notice was that all, repeat all, the bullet holes were in the Parti Québecois side of the parliamentary chamber. That fact was never mentioned in the news reports.

They were his target and they weren't there.

Lortie shot his way to the speaker's chair and sat down.

There was a video camera mounted over the door that Lortie and we had entered.

Sitting in the chair Lortie saw the camera. He was angry and frustrated. He had come to Québec City fully prepared to kill the members of the government in power. He had failed.

He again sprayed the camera with bullets from the speaker's chair. He missed.

The camera silently recorded his every move and the microphones picked up and recorded every word and shot.

The tape for the recording camera had been changed minutes before the scheduled commencement of the parliamentary proceedings, so the whole time Denis Lortie was there it was all on the same tape.

We were taken to another room and Lortie's packsack was sitting on a desk. I asked if I could look and the cop just shrugged and said go ahead.

The military packsack had a gas mask, first aid kit, and several empty and two more full Sterling machine gun clips, plus several full clips of 9 mm rounds for the pistol he carried. We went back and looked at the parliamentary chamber strewn with 9mm shells and the blood trails from some of the 13 people who were wounded and removed to nearby hospitals.

One hell of a day, go to work in the provincial parliament building and come home with a repaired bullet hole in your body, or dead.

A few minutes later, we were all called to a large conference room where the tape was to be shown to all the cops who were present.

The whole episode lasted a little under two hours. Fresh, uncut with no editing at all.

I was the only person in that room that was not wearing a badge and a gun.

Denis Lortie was clearly very angry that his targets were not there.

We heard the whole conversation between Corporal Denis Lortie and Sergeant-at-Arms Rene Jalbert.

The event was a spectacular news story but the tape of what had happened was not aired on TV for a year, and then, only a few seconds with Jalbert talking to Lortie. No photos of the interior damage were ever aired or printed. Lortie's packsack was never mentioned.

When we walked out to the main entrance we passed through the anteroom where the three people had been murdered. The pool of blood had been cleaned from the glass tabletop and there was a worker in there with plaster and a spatula filling the bullet holes in the walls. He also did the plaster walls in the main parliamentary chamber, but the bullet holes in the wooden desks are there to this day, but I suspect they were filled with plastic wood. Security is now very tight for anyone entering the parliamentary building. Politicians always look after themselves first. In this case with reason, just too late.

We left and went back to the cop shop and to the photography department. There I took one of the aerial photos that showed Lortie's car by the side entrance and printed a black and white copy. I took no interior photos at all. I took that aerial photo to the newspaper and sold it to them. They had the only aerial photo of the parliament building and the cops did not mention how Lortie had arrived and did not describe his car.

The newspaper bought the photo. I charged them $100.00 and gave them a receipt from "Photo Roy".

Remember "Arrow Why" from an earlier chapter, or seal? Again, I was playing with French. Roy, in French, is a family name, in English, it is a first name, like my father's first name. Roy is spelled r-o-y in English but pronounced Arrow Why like the game warden wrote on my dad's fishing license. Roy is pronounced "rWa" in French. It was in a French city, in a French newspaper, and nobody would connect that to an English name.

The next morning the newspaper was published with the spectacular story of the parliamentary murders and the aerial photo of Lortie's car on the south side of the parliament building, with "Photo Roy" printed at the bottom of the photo. It was an exclusive bit of news for the paper and they took full advantage of it. The photo was enlarged to cover the top two-thirds (2/3) of the front page.

No one ever questioned me how the newspaper got the aerial photo and the information that it was Lortie's car in the photo, or the backpack and its contents. The cops didn't know that I had declared an exclusion zone around the parliament buildings, which was between me and the control tower. The cops never listened to me talking to a controller on an aviation frequency, and I could control that with the audio panel switches in the helicopter, and I never let my observers listen to the aviation frequency. Not only that I always communicated in English on the aviation radio frequencies. It was easy, two different languages, two different radios, that I controlled.

The photo cost me nothing. I didn't pay for the film, the development, or the printing of the copy.

Their newspaper editor was very happy, he got an exclusive photo, an exclusive bit of information about the massacre, and he took advantage of everything, and sold a lot more newspapers than normal.

Float plane thief

I was called to go north to Chicoutimi, in the Lake St Jean area.

Several cabins in the area had been broken into, and anything of value was missing. There are hundreds of small lakes north of Chicoutimi, and many have cabins on their shores. They are excellent for fishing and moose hunting. These were cabins that had no roads to the lakes, and the only means of access is by float plane or snowmobile in the winter.

I went with a cop to visit several of the cabins that had been broken into and had stuff stolen from them. Some of them I couldn't find a place big enough to land so I would have to let the cop off on a small dock or place on the shore where I could get close enough without hitting the trees with the rotor blades.

We visited several of the cabins that had stuff stolen from them, and in all cases, they did not have road access. After a long day, we were flying back to base and saw a couple of cabins that did have road access, and one had a second cabin further back in the bush. This cabin had several canoes on racks beside the isolated cabin. The cop took note of it. We landed at the dock of the cabin by the lake and the cop took a walk back to the cabin with the canoes deeper in the woods. He couldn't get in, but he could see a lot of camping gear through the cabin window. In particular were several outboard motors.

He didn't break into the cabin, but he could get a search warrant.

We finished our day, and I headed back to Québec City.

The next time I went up to Chicoutimi, I sought out the same cop and he told me that he had gotten his search warrant and that several of the motors he had seen had their serial numbers checked and were stolen.

The owner had been arrested and did own a small float plane.

Many of the items found in the two cabins had been recorded as stolen.

The story hit the news, and a couple of months later, when I went back to the area, both cabins were burned to the ground.

I never knew if his plane had been damaged by his irate victims, but he had been caught, charged, and prosecuted. His victims took their own revenge.

Tall Ships

The tall ships arrived in Québec.

They are top sail schooners, schooners, ketch, sloop, brigantines, brigs, and barques or barks. They are monohulls, catamarans, and trimarans. In all, there are more than 100 tall ships in the world. They are used as school ships, museum ships, restaurant ships, and cruise ships.

Some were three-masted Barques or Barks and several two-masted schooners, like the "Bluenose II" from Nova Scotia. The "Bluenose II" is famous in Canada because not only is it based in Lunenburg, Nova Scotia, but it achieved immortality when it was engraved on the Canadian .10₵ piece. A twin-masted schooner, it was a fishing and racing ship built in 1963. The original Bluenose was also built in Lunenburg in 1921, struck a reef, and sunk off Isle aux Vache, Haiti, on 28 January 1946.

The tall ships went to various places around the world to be shown off and displayed. They charged for a tour of their ships and brought in many tourists.

The tall ships filled the Québec harbour and were a great attraction for the city. I was lucky to be able to fly around them with the police helicopter and got some great pictures.

I had landed at the cop shop for a while, and while shutting down, a man came to the helicopter and showed me his ID as an inspector for the Minister of Transport. He was arrogant, like some with power, and demanded that I show him my documents, all the documents.

There are 10 things that a pilot must have in his aircraft with him. Pilot's license with endorsement on type, current medical, C of A (Certificate of Airworthiness), C of R (Certificate of Registration), Current Canada Flight Supplement, and 5 other documents. I asked him why he was checking a government-owned aircraft. His answer was lame and arrogant. Hoping not to see him again I left after having shown him the documents. I was wrong.

I finished flying over the tall ships and, from my base, went straight to the bank to get some money. I was still wearing my uniform, but I had to stand in line just like everyone else.

This was before the days of our now-familiar ATM. Automatic Teller Machine. The bank did provide a limited-time teller service after banking hours. From closing time and for about three hours there were two tellers in heavily protected cubicles to service their customers. As I was standing in line there were two American tourists and several people in front of me. I could hear them asking one of the tellers to exchange some US dollars for Canadian dollars. They couldn't speak French and the teller was making her understanding of what they wanted very difficult for them. She refused to make the exchange telling them they were not customers of the bank.

They turned away from the teller, dejected and unsatisfied with the teller's refusal to make the exchange.

The teller pissed me off.

The downcast couple was about to pass by my side when I touched the man on his shoulder and told him in English to wait a few minutes. I smiled at them and said that I would handle it.

They waited about five minutes until it was my turn to talk to the teller. I showed her my bank card, and then I wrote a check for cash from that bank to myself. After taking my cash, I was finished with what I wanted.

I then raised my voice and told the teller that she was stupid and incompetent. When a person makes an exchange from one currency, in this case, US dollars, to another currency, Canadian dollars, the banks charged a fee for that service. They made money. The bank paid her salary so she was depriving the bank of money to pay her. This dissertation I made in a loud voice and in French. Everyone around me could hear and understand me.

I turned to the American couple and, in English, asked the man to give me the money that he wanted to exchange. He did, $300.00 US.

I turned back to the teller and again, in a loud voice, told her that I was a customer of the bank and I wanted to exchange $300.00 US for Canadian dollars. She had no choice but to comply.

With the transaction complete I gave the Canadian money to the American man and told him to enjoy his visit to Québec City and that not everyone in Québec was nasty and arrogant. After they thanked me with a handshake, we parted.

I was quite happy with my "Good Samaritan" action for the day.

The next day there were thousands of people on the docks wanting a tour of the tall ships. I was admiring the majestic ships from one of the best places to view them, in the helicopter.

We received a call to land on the Plains of Abraham beside the old fort or citadel. A passenger was aided into the back seat by my observer. It was the MOT inspector who had been so arrogantly demanding the day before. I tried not to be very cooperative, but I also wanted to get rid of him as soon as possible.

He was looking to apprehend a person who was using a hang glider to observe the tall ships. He would take off from the Plains of Abraham, glide around the tall ships, and then land along the highway to be picked up by his accomplice and then come back up the cliffs to do it again. Minutes later, we spotted him over the tall ships gliding back to the shore along the highway to be picked up. We followed him down, and I dropped off the arrogant MOT inspector. I do not know what happened, but I do know that the hang-glider pilot was in for a rough and expensive time.

That summer was busy with several searches, another bank robbery, and there was a game warden who had been shot at by a couple of poachers. He was unhurt, but after that, game wardens worked in pairs in that area.

The meeting:

In the middle of the summer, I was at the office one day, and a flight dispatcher called me over and told me a representative of Bell Helicopter was here and wanted to talk to the chief pilot. The chief pilot, Ben, was not there and Ben couldn't speak English anyway, so the dispatcher referred the Bell rep to me. With his accent, he was obviously from Texas, Bell's head office. I led him to Ben's office and we sat down for a chat.

We discussed the small fleet of Bell 206B's that were owned by the Québec Air Service and did we had any intentions of buying more machines.

He then asked me what an English guy was doing living in Québec City and working for the Government of Québec.

I told him that I had been in the military, and they liked that, so they hired me.

"What did you do?" he asked.

I told him that I was an ex-US Army and a Vietnam Veteran and flew Huey Cobra gunships in Vietnam for a year.

"And," he asked.

I then told him that, using the GI Bill that, I went and got a US ATP, IFR, and had a current FAA medical on top of my Canadian license.

He looked at me, eyes wide, and said show me. I took out my wallet and showed him my US ATP and my current FAA medical. "Give me your resumé," he ordered. He told me that Bell Helicopter is going to build a manufacturing plant north of Montreal to build all of Bell's commercial helicopters.

He gave me his business card and told me to mail my resumé to him.

The Pope

The Pope was scheduled to make a tour of Canada in mid-August. He had a scheduled stop in Québec City. The Pope had been shot and seriously wounded a couple of years before, so the security of the Pope was a major undertaking. After a couple of days in Canada, he arrived in Québec City. His "Popemobile" was airlifted into Québec City with him on another aircraft.

The police issued a lapel pin to every member of his security detail. No person without a lapel pin could even get close to him, and that included the people who handled the baggage, plus the maintenance and refueling people who got near the Pope's aircraft.

I also had a lapel pin; in fact, I still have it all these years later.

The Pope passed within arm's length of me as he was led to his armoured car.

He was taken downtown to meet with church leaders, other dignitaries, and politicians.

The next day I was scheduled to provide aerial security for the Pope and his entourage on a visit to the very famous Basilica of St Anne de Beaupré 19 miles (30 km) east of Québec City.

The highway had thousands of people lined up along the way to catch a glimpse of the Pope. The caravan of vehicles, with the Popemobile in the center, slowly moved along the highway. The people were cheering and waving to the Pope as his motorcade made its way to St Anne de Beaupré.

The Basilica of St Anne de Beaupré has, for over 350 years, received millions of pilgrims. It is the oldest pilgrimage site in North America. This Notre-Dame Basilica is also named as one of the most beautiful buildings in the world.

Every Thursday there is a mass celebrated that is dedicated to St Anne, the grandmother of Jesus.

This shrine is also well known for healing thousands of its pilgrims. There are, in fact, thousands of canes, crutches, and wheelchairs lining the inside walls of this ancient structure. Over 11 million people visit this Catholic shrine each year.

My helicopter was the only aircraft that was allowed anywhere near the Pope. The military provided perimeter security, but even they were not allowed closer than 5 miles.

I remained in constant contact with the Québec City control tower and they kept me advised of any traffic anywhere near me.

(A NOTAM, Notice to Airmen, had been issued by MOT prohibiting any unauthorized flight within

5 miles of the Pope's location, which was specified in the NOTAM)

I had the sky to myself and my camera anywhere near the Pope. And I took full advantage of that opportunity.

The Pope, after arriving in St Anne and touring the Basilica had a mass scheduled to be given from the front steps of the church. There had amassed several thousand people to hear the Pope over the dozens of loudspeakers set up outside the Basilica.

I could not stay too close because of the loud noise signature emitted from the helicopter, so I shot a couple of pictures and moved away for the duration of the Pope's sermon. This was also the time to refuel the helicopter. His sermon would last at least an hour and I would be informed by police radio well in advance of the Pope's departure.

The timing worked well; I was back over St Anne de Beaupré about 10 minutes before his motorcade departed the Basilica for the return trip to Québec City.

We again supplied aerial security for the Pope. After the Pope's motorcade had returned to where the Pope would be spending the evening and the motorcade had dispersed, I went back to the airport.

I filled out the aircraft logbook and then went home. There I changed out of my uniform and went to the cop shop photo department. The cop on duty helped me develop the 3 rolls of film I had taken, and I then picked the 8 best photos and went to the newspaper office. There, they picked 3 photos that they wanted to print. Pages 1, 2, and 3 of the newspaper each had a different photo. They were all above the fold and each took half a page or more in the next day's newspaper.

I charged them $200.00 each, far more than my regular price. The editor didn't blink an eye and paid me cash. The editor told me he was going to double the run the next day for the paper. He was right, all copies sold out. Again, the photos were supplied by "Photo Roy". Again, nobody said a word; nobody "clicked." The only person who knew it was me was the cop who helped me develop the film, and all he wanted was a couple of copies of a few of the photos.

The next day, the Pope was to take his motorcade through both lower town and upper town of Québec City. From there he was to proceed to the University of Laval to give a sermon in their football field. We again escorted the Pope's motorcade everywhere, and when the Pope arrived at the University, I took a couple more photos. The huge crowd was impressive and made for a couple of great pictures. Then, it was back to base to refuel. With time to spare, we were informed of the Pope's departure, and we were again overhead for motorcade security on his trip back to his overnight quarters.

I had bought a telescopic lens for my camera and used that as justification for the higher photo price. Again, the editor paid for a couple more photos and they were printed above the fold, taking half to two-thirds of the page. The newspaper editor again doubled his run and sold out.

Each photo printed was credited to "Photo Roy." Those photos paid for my new lens and the cost of the camera. I explained before about "Photo Roy" with my Denis Lortie photos. Nobody ever questioned where the pictures came from, even though I was the only aircraft within 5 miles of the Pope.

The next day, the Pope took the train to Trois Rivières, but it was raining hard with low overcast clouds. I left him at Trois Rivières and only got one good photo, which was again printed.

The next day, the helicopter from Montreal picked up the Pope's security cover and I was finished escorting him.

I was curious, and I tried to find out about other newspapers, yet no other newspaper in Canada printed an aerial photo of a Pope's mass or his motorcade.

The last tree

This was my last week of work for the Sûreté du Québec and the Québec Air Service.

I got a call to go east near Rivière du Loup (Wolf River) for a search. Another fisherman was lost.

I had to pick up the dog master and his German Shepard before going to the search area.

Normally, I didn't try to hurt his feelings, but today I just didn't give a damn. I told him that after we picked up the local SQ officer who was to show us where to look, I would drop him off, fly around a bit, and the lost person would build a fire while I was gone to Rivière du Loup to refuel. I'd find him when I came back.

He got insulted, turned around in his seat, opened his packsack, took out his coveralls, and put them on in the front seat. He then stretched back and put the harness on the dog. He was ready to go when he could open the door.

We got to a small church a few miles before Rivière du Loup and picked up the SQ officer as my observer.

The story was about a lone fisherman who had parked his truck not far from a small river the morning before and had not come home.

It was a beautiful, cloudless fall day, bright sunshine and not a breath of wind. Totally dead calm.

We headed southwest to a large valley going south. From the hills south of the St Lawrence River, the small river opened up to reveal a large flat valley. The SQ officer pointed out a sugar cabin on the west side of the river. To my left was the widening valley.

My peripheral vision caught a slight shimmer in the calm air down the valley. A column of rising heat had passed on my left between me and the sun. It happened too fast to see when I turned my head to focus my eyes on it. My 6th sense developed over 11 years of searching had seen the heat shimmer.

I knew where the lost fisherman was even faster than I had told the dog master before on our way here.

The fisherman was half-smart. He had built a fire, but he made it with very dry wood. No smoke, only heat. He didn't put anything green on the fire to make smoke.

The totally calm air let the heat rise, but only the brief shimmer while passing the sun was visible to me. That was enough. All the years of training my mind to see using my flying 6th sense had again been worthwhile.

A single breath of wind or a cloud covering the sun would have negated the effect, and we would have had to spend a lot more time searching.

I told the cop that I had found the lost fisherman. He looked at me in disbelief, and then I asked him to show me where the fisherman had parked his truck.

I flew directly to the truck and then, at a low level, turned towards where I knew the fisherman was waiting. The lost fisherman had heard the helicopter and knew we were close.

I passed directly over him at a low level and rolled left to let all of us see him.

There was a small swamp about 100 yards away, and I dropped the dog master and his dog off and waited for him to return with the fisherman.

We all returned back to the church and I dropped off the local cop and the fisherman. He would be taken to the hospital for a quick check and then to the cop shop to record his story. Getting his truck back was his problem.

On our way back to Québec City, the dog master was very quiet.

I dropped him off, knowing I would never see him again.

My career as a police pilot was ending the next day.

A week later, my wife, daughter, son, our dog, and I boarded an airplane for a new life in Fort Worth, Texas. I was to go there for two years to be trained as a test pilot for Bell Helicopter TEXTRON, one of the world's top helicopter manufacturers.

In the 11 years I worked for the Sûreté du Québec, I flew 5,222 hours or an average of 475 hours per year, which was actually 6 months with the 7 on, 7 off schedule. 475 hours divided by 6 months is 79.1 hours per month, which is almost exactly the maximum number of flight hours that an airline pilot captain is allowed to fly, but with a co-pilot and an autopilot to help him.

Flying for the police was exciting and extremely satisfying, but now I would be taking a new road in my life's journey.

It seemed that every job I had or was to have had been judged to be the most dangerous for a helicopter pilot. Combat pilot in a war zone, a police pilot constantly putting myself in dangerous positions to save the life of somebody, and then as a helicopter test pilot, finding the performance boundaries of a new model helicopter to set the limitations for the use of that model helicopter for commercial use. Also, as an instructor pilot where, there were always a few who would try to kill you, and yes, that did happen on a few occasions. I learned to be a very careful and aware pilot. Shucks, why not? I'm still alive to write this book.

PART 2: Particularities of a helicopter.

A helicopter is unlike any other type of machine and the aerodynamics of a helicopter are sometimes difficult to understand.

I was taught about settling with power but never why.

I was told about retreating blade stall, but never the physics of precession that affects it.

All this and much more I had to learn and explain to myself.

Much of the following I was taught, but much was poorly explained, and there was never a book or any teaching material that I ever saw in 47 years of flying that explained or taught all that I have included in this book. As a Bell test pilot every helicopter MUST fly a minimum number of hours to be certified. One reason is the low fuel light. That leaves every pilot testing new helicopters with many hours to fly to burn all those hours. That's when I took the time to teach myself everything I'm putting in this book. I was just curious and when something happened, I wanted to know why, and I had many hours of flying to learn just that.

With Bell, I flew every model they ever produced up to 2005, except the V22 Osprey.

If you are a pilot, mechanic, aeronautical engineer, or anybody and disagree with anything in this book, please contact me at my email address at the end of this book.

The helicopter: The following information will be described using a Bell 206B JetRanger or the Bell 206L series LongRanger.

I will add helicopter characteristics at various times as I write this book, but first, I will explain that a helicopter cannot be a helicopter if it does not have a swashplate.

The swashplate was invented and patented in June 1909 by Jules and Paul Cornu in France.

Harold Pitcairn and Juan de la Clerva are credited with developing the swashplate. The first helicopter was the VS-300, patented by Igor Sikorsky in 1939, a Russian who had immigrated to the USA, while Lawrence Dale Bell built his machines at about the same time. Both their machines started being used in World War 2.

There are still 2 large helicopter companies bearing their names.

Translational lift:

The wind is a major factor during the operation of a helicopter in a hover.

What is a hover? A hover is the toughest thing to do before becoming a helicopter pilot. It is the ability to hold the helicopter over a single point over the earth for a period of time while controlling all the forces that are trying to prevent the pilot from doing just that.

1) Torque from the engine tries to turn the helicopter opposite the direction of rotation of the rotor blades.

2) The lift from the rotor blades is variable depending on engine power and the weight of the machine. Normally, on turbine-powered helicopters, a governor will keep the rotor speed constant throughout all realms of flight. Piston engine helicopter had the pilot roll the throttle on as he applied power. Essentially, it was up, on, and left. Up collective power, increase power on the throttle, and push the left pedal to counteract the increased torque.

Plus, at the same time, the pilot must continue to maintain hover equilibrium at a hover with the cyclic control stick no matter the force or direction of the winds affecting the helicopter.

3) Engine power is determined by the weight of the helicopter and the size of the engine.

4)   Wind speed will constantly change the direction and the speed of the air across the rotor disc.

5)   Torque will constantly change due to the power pulled, wind speed, and, therefore, the pedal input of the tail rotor pedals. The pedals will constantly need to be adjusted to the constantly changing torque from the engine. If you consider zero wind, then the inputs should be constant and, therefore, zero. Tail rotor pedals control the rotation of the helicopter. The power pedal (left pedal in a Bell 206B) or when looking down at the rotor disc, the rotor blades move counterclockwise, requiring the use of the left pedal. It's the power pedal because it increases the blade pitch angle on the tail rotor blades and that causes an increase in torque, or power from the engine. The tail rotor prevents the nose of the helicopter from turning right due to the torque produced, which is why they can also be called anti-torque pedals.

The bottom line is that hovering is not easy, and if you cannot do it, then you will never be a helicopter pilot. Hand-eye coordination is where you make it. A gymnast may never be a helicopter pilot if he does not have hand-eye coordination, but a piano player may make a good helicopter pilot even if he can't walk and chew gum at the same time.

Oops, one problem: if he joins the military, he won't make a good soldier on the drill field. My good friend, roommate, & the class valedictorian was such a man. The second person to solo on my flight was me, with less than 12 hours of helicopter flight time. The first was a commercially rated helicopter pilot who joined the US Army; the third was George, who had never flown anything bigger than a kite. A gymnast who was a near Olympic wonder never even got close to learning how to hover. Where is the line, where is the ability, only the human mind can open the road to what ability is possible.

The hands and feet don't fly the helicopter. They are only the tools the mind uses to control the machine. Situational awareness is the key to your understanding of where you are and what you are doing in three dimensions.

At the beginning of the Vietnam War, the US Army soon realized that they had to have thousands of helicopter pilots to make the Vietnam War work for them.

The Vietnam War became the first helicopter war.

They only had a small pool of experienced helicopter pilots to draw from. They decided to enlist pilots as Warrant Officers and train them to be helicopter pilots.

That idea was a spectacular success. Smart 18-year-old teenagers were brought into the US Army and trained as helicopter pilots. They were young, easy to train, invincible, and produced a generation of helicopter and airline pilots that became the backbone of the growing airline industry.

The VHPA directory lists thousands of ex-Vietnam helicopter pilots who used the GI Bill to get their fixed-wing Airline Transport Pilot License and became airline pilots.

Understanding the wind can be very useful if you can use the effects of the wind to your advantage.

Hovering into wind with a 5 or 10-knot wind will produce visual cues close to you if you know what to look for. Pilots will normally search much further than they need to, to find the wind direction. The most obvious is, of course, the windsock. The leaves on trees are very good indicators because the bottoms of the leaves roll into wind. Leaves rolling into wind actually produce lift for the branch, which will reduce the drag from the branch creating the resultant drag force sufficient to overcome the natural bending of the tree. Smoke, grass, water, and any other thing that shows the effects of wind could be your guide.

The closest and often most helpful indicator is often ignored or not known. The surface below & in front of the helicopter is often the best indicator of the wind direction.

The dust, the sand, the grass, or other light refuse on the ground that is blown up by the rotor wash when at a hover will tell the pilot exactly from which direction the wind is blowing. The added bonus is that these indicators will also tell the pilot exactly where he will enter effective translational lift. This is also the time when FOD (Foreign Object Damage), is the most likely to cause damage to the helicopter.

I will call the point at which the rotor wash, evidenced by the debris lifted by the rotor wash around the hovering helicopter and clean air meet, the "wall." This wall, if studied closely, will have the disturbed surface of the rotor wash moving, lifting, and disturbing the area around the helicopter by the power of the air generated by the rotor blades.

Look at the helicopter from directly above from about 300 feet over a grassy field. Assume a wind speed of 10 knots. With the helicopter nose pointed directly into the wind, the rotor wash disturbing the grass below the helicopter would look like a pear.

The distance from the nose of the helicopter to the disturbed grass would be the shortest around the diameter of the disturbed grass by the rotor wash when the nose of the helicopter is directly into the wind. On the two sides of the helicopter, the disturbed grass would bulge out like the sides of a pear. The rear of the helicopter would have a long area of disturbed grass that would become indefinite as the distance from the rotor of the helicopter grew greater.

The often-taught point of translation lift is 15 knots. What happens if the pilot elects to take off crosswind? What about a take-off on a windy day over an obstacle? All these variables can be worked with if a clear understanding of translational lift is known.

The nose of the helicopter should be pointed towards the flat part of the pear-shaped disturbance. Sitting at a hover at about 3 to 5 feet above the ground, facing into the wind over the grass, the pilot would accelerate slowly and smoothly into the wind. The previously explained "wall" would approach the helicopter, and when the mast of the main rotor passes through this "wall," the helicopter will have entered effective translational lift.

At this point, the helicopter will immediately start to climb. The pilot of a Bell 206B helicopter will be able to lower the collective pitch by 8% torque and still maintain the climb, in fact, if the collective is not lowered the torque will decrease by 8% by the time the airspeed is 60 knots.

Effective translational lift will occur during take-off when the mast of the helicopter passes over the point where the disturbed air and the mast of the helicopter meet. This point of effective translational lift will be the same for every model and type of helicopter. The direction of rotation of the blades, the speed at which the take-off is made, or the steepness of the departure angle are all the same as far as the point of effective translational lift that would be attained.

When at a hover, each rotor blade creates lift, one at a time, to overcome gravity. Effective translational is the point at which the rotor blades become a rotor "system." The rotor disc acts like a plate, a dinner plate in flight, or like a Frisbee. The rotor disc as a whole becomes the power, like the wing of an aircraft, instead of the individual rotor blades being the force.

So, to add it all together, what I have described is true of all helicopters, always. The place where the disturbed and undisturbed air meet at the mast of the helicopter, no matter what the wind speed, size of the helicopter, the weight of the helicopter, take- off profile, nor the direction of rotation of the rotor blades, will make no difference, the point when the mast passes through the disturbed air into the undisturbed air is where the helicopter will enter the effective translational lift.

There are three variables when entering translational lift:

The speed of the departure, in effect, the angle of attack of the rotor disc, will reflect the required power needed for the take-off.

If the helicopter is moved forward slowly there will be a dip of 2 to 4 feet just prior to entering translational lift in a low wind condition.

The direction and speed of the wind are in relation to the direction of take-off.

In a slow, careful take-off, with the pilot keeping the helicopter within the limitations of the height-velocity curve printed in the flight manual in the "Limitations" section of a single-engine helicopter or the "Performance" section of a multi-engine helicopter, is a safe take-off. The reaction to an engine failure during a power change or a high-power situation would require a far less aggressive recovery technique.

What about landing?

The reverse is true. Should the pilot decide to do a circuit from the take-off point, the pilot would note the power required to hover. The pilot could take off, do the normal circuit, make an approach, and while looking to his right (Helicopter Captain on the right-hand side) during the approach, the pilot would see the rotor wash disturbing the surface below him/her approaching from the rear and as the disturbed air passes forward through the position of the mast the pilot would pull the previously noted hover power and level the helicopter for a landing attitude and the helicopter would settle into a hover at the same height above the ground. To make the landing smooth, pull the collective power as soon as the disturbance is even with the mast. The transition from translational lift to hover will be smoother and less aggressive. One other point to note is that when the pilot pulls collective pitch in anticipation of the power required to hover, the pilot should add a power pedal application equal to the required pedal input needed for a hover. This, again will make the hover a smoother maneuver. One other point to note is that if a right turn after the hover is anticipated, do not add left pedal or power pedal until after the helicopter starts to yaw to the right. Then, the pilot would stop the yaw with the pedal. Again, this makes the whole maneuver require less power, make it smoother and feel better.

This is especially true in snow, dust, or other obstructions to vision. When you know that you will be landing in an area of potentially poor visibility, pick a rock, post, tree, or as a last resort, throw anything that would not be blown away by the rotor wash or something attached to a length of trail tape out the window on a low pass prior to landing, and use that as your guide to your point of intended landing. This will give you a point of reference, wind direction, and a fair idea of the wind speed. Never, and I mean never, ever even try to land someplace where you do not have a positive point of reference. There will be two choices: land safely with a positive point of reference or crash. You may get away with landing without a reference once or twice, but you will bust your machine if you continue your lack of planning.

Magnetic variation:

Magnetic declination or variation is the angle between magnetic North and true North. Declination (variation) is positive when this angle is east of true north and negative when it is west. Magnetic declination changes over time and with location. For example, Montreal is 13.7 degrees west. To fly True North on a map you would need to fly 13.7 degrees. If a pilot were to fly 0 degrees magnetic in Montreal, he would actually be 13.7 degrees west of true north. A magnetic variation on a map is either in the legend or on a compass rose printed on the map. Magnetic variation in New York City is 13° west, and Los Angeles, California is 13° east. Magnetic Variation in St Louis, Missouri, is essentially zero.

Magnetic north

I was curious, and I tried to find out about other newspapers, yet no other newspaper in Canada printed an aerial photo of lifetimes, but it has changed poles, and it will happen again. Human life could not exist on planet Earth without a magnetic field protecting us from cosmic radiation and charged particles from our sun.

Settling with power

The helicopter increases its rate of descent with an increase in power or collective pitch. This will happen if the machine is not in translational lift and is usually downwind. The airspeed indicator would show zero speed.

When this happens, the only way to recover is to lower the power, (collective pitch) and the nose of the helicopter. As soon as you re-enter translational lift, the airspeed indicator will show a positive speed, and then the pilot can pull power and re-enter forward flight. This will happen fairly quickly and the pilot must react fast.

The pilot will be required to sacrifice some altitude to do this, and if this condition occurs with the helicopter too close to the ground, he will crash the machine, like the example of the V22 Osprey in the introduction of this book.

The one mentioned was the third V22 crash. The

first crash was at the Boeing plant, where the Boeing fuselage was mated to the Bell wing with the engines.

At Bell, the aircraft are always tied down so that maximum power could be applied and the aircraft would not move. Boeing did not want to do that.

They also screwed up the mating.

A Boeing helicopter test pilot was at the controls, and when he applied power to the twin rotors, the aircraft lifted into the air. He tried to control the aircraft, but the aircraft is fly-by-wire, meaning the control actuators are electrically driven, not with metal control rods. The fore and aft control wires were properly connected, but the left/right wires were crossed. He noticed the discrepancy right away and pushed the collective down, oops, the power lever, nicknamed the Blottle, instead of the throttle after the Marine project commander, Colonel Blot. The aircraft went up.

The pilot lost lateral control and wanted to go down and, with the instincts of a helicopter pilot pushed the power lever down. When the V22 went up, he had to reverse his move and pull the power lever up. By this time, he had about a 30° left roll, and when he went down, he crashed. The 2 errors were kept quiet.

Colonel Blot wanted to attract Harrier pilots to fly the V22 Osprey. He refused to consider that most fighter pilots of the Harrier would be little inclined to transition to a flying truck. The V22 Osprey can carry 24 seated troops or 32 when standing, or internal cargo or the cargo hook could be used for loads up to 8,300 lbs (3765 kg).

The design of the power lever (Blottle) was changed to make it more like a fore/aft throttle movement in the Harrier jet.

Harrier pilots want to carry rockets and bullets, not people or groceries.

(The Harrier jet was featured in the movie "True Lies" with Arnold.)

The second V22 Osprey accident happened in Washington, DC. The machine was preparing to land, and the engine nacelles were tilted up in preparation for landing. Unfortunately, there was a leak from one of the proprotor gearboxes and some oil had had a chance to leak into the intake of the engine. When the proprotors were tilted up, the oil was pulled into the engine, started a fire where there was no suppression, and the aircraft crashed into the Potomac, killing all aboard. The discrepancy was found, and thereafter, a hole was drilled in the intake to drain any oil accumulation.

All 3 accidents were caused by human error. The

V22 Osprey has since proven one of the best, most versatile, safest, and most successful purchases by the US military.

Tail rotor and take-off:

The tail of a Bell 206B helicopter has a vertical fin installed with a 5-degree offset to the right.

The pilot should start the take-off from a steady state hover at 3 to 5 feet above the ground.

The idea here is to use the minimum torque for take-off. If the pilot were at a 5 to 6-foot hover and consciously not moving the collective during the take-off, the pilot would notice a 2 to 3-foot dip in altitude as the helicopter approaches translational lift. Passing through translational lift the helicopter would continue to accelerate, and the helicopter would also start climbing. The pilot would continue the climb attitude and continue not to touch the collective. The helicopter would accelerate through 60 knots, and at that point, the pilot would check the torque gauge and notice that the torque would now be 8% below the power required to hover.

The vertical fin with the 5 degrees of right angle will have taken over the tail rotor power that was required to hover. As the helicopter increased speed, the pilot would have slowly decreased the amount of left pedal power applied. The idea is to keep the helicopter in a level-trimmed attitude.

That means that the "ball" is centered.

In a helicopter or an airplane, the amount of yaw force is felt by the amount of yaw pedal input that is applied. This force is a lateral "G," gravity force.

In a car going around a sharp curve to the right, the lateral "G" force would push the driver and the passengers to the left side of the machine. The faster you go, the greater the lateral "G" force. There are three things that greatly affect this force, the lowest possible center of gravity of the vehicle (a high center of gravity would give the vehicle a tendency to roll over much more easily), the tires that touch the surface of what they are riding on, and the surface itself.

In an airplane or a helicopter, this lateral "G" force is the amount of pedal input by the pilot in straight and level flight.

The Bell helicopters also have one other aerodynamic "fix" to reduce the amount of power pedal input required, and that is the vertical fin. The vertical fin or other vertical surfaces on all turbine-powered Bell helicopters are shaped like a wing turned 90 degrees and then installed on the tail of the helicopter to reduce the required amount of tail rotor power needed by the helicopter when in flight.

The tail rotor in a hover pushes the air to the left of the helicopter, counteracting the torque of the main rotor blades.

The vertical fin in flight also pulls the tail of the helicopter to the right to counteract the torque from the main rotor blades.

In all Bell helicopters in level cruise flight at maximum continuous power, the tail rotor pedals should be even with each other if rigged properly.

In helicopters, where the main rotor blades turn clockwise when seen from above, the power pedal is the right pedal.

Then there are the helicopters with no tail rotor. These have two main rotors turning in opposite directions, and the torque is applied to the fuselage as a twisting motion. The CH47 "Chinook," widely used by the US Army, is an example of that type of machine.

One other type, though rare in the USA but used by the Russians, is the counter-rotating main rotor-bladed machines. Also used on some of their airplanes are the counter-rotating engine propellers, which are very noisy. The Western countries don't use counter-rotating propellers because of the excessive noise, whereas the Russians don't care about human factors.

Autorotation or quick stop:

From a hover, an aborted take-off can be for two reasons: A) an engine failure, or B) a quick stop.

The forces on the rotor are mechanical in a quick stop and aerodynamic during an autorotation.

During a quick stop, the engine drives the transmission, which turns the mast, which rotates the blades, keeping them in the same plane of rotation, thus forcing air down onto the horizontal stabilizers, forcing the nose to lift up, increasing the flair.

During an autorotational flair, the driving force on the blades is aerodynamic. The force turning the blades comes from the air moving up through the blades. This force results in the coning of the rotor blades, which when coning, does not force air down onto the horizontal stabilizers, lifting the nose, but rather lowers the nose of the helicopter in a levelling action.

Autorotation:

During an autorotation with power off, the nose of the helicopter will pitch down when the collective pitch is pulled.

With power off, the nose-down pitch is caused by the coning of the rotor disc, which causes the aft portion of the rotor disc to flap up, causing the pitch down of the nose during an autorotative flair. A flair or a turn causes increased inertia, and the rotor blades will increase their RPM. Care must be taken to prevent the rotor RPM from exceeding the upper red line limit. With an engine failure on take-off, the pilot wants to keep as much rotor RPM as possible. With the tendency of the machine to nose down the pilot must apply aft cyclic to keep the nose up to reduce airspeed, maintain a high rotor RPM and sufficient altitude to slow the helicopter as much as he can. The pilot will need to pull some collective pitch to stop the forward motion, still keeping the rotor RPM in the green range, then lowering the nose to a level attitude, and pulling the collective pitch helps that. The flight manual suggests that the pilots plan to put the helicopter on the ground before 70% rotor RPM. The pilot is wasting over 10% of rotor RPM if he does this.

To test this, hover the helicopter at about 2 to 4 inches above soft ground.

Cut the power and hold the skids of the helicopter at 2 to 4 inches above the soft ground. In a Bell 206B, the helicopter will fall from the sky at 56% rotor RPM.

That's 14% lower than the flight manual suggestion. Bell's safety margin is for the lawyers, but a knowledgeable pilot can use that RPM in a real situation to get the machine safely on the ground at zero or near zero forward speed.

A careful helicopter pilot of any make or model would want to test his particular machine to find the reality of the rotor rpm of his helicopter over soft ground. From a maximum of 2 or 4 inches, the machine will drop like a rock with the loss of lift.

Hovering Autorotation:

I will describe this as it happens in a Bell 206B JetRanger or LongRanger with counterclockwise rotating main rotor blades.

A hovering autorotation is normally practiced at 3 to 5 feet above a flat surface. The goal is to simulate an engine failure at a hover. The greatest percentage of turbine engine failures occur during a power change. Lifting to a hover is the first major power change after starting the turbine engine.

The last major power change in any flight is the approach to a hover prior to landing.

Practicing the hovering autorotation is quite simple but is complicated in what is required for a smooth hovering autorotation.

First, from the established hover, the pilot would chop the throttle from 100% power to flight idle at 61% turbine speed. The rotor RPM at 100% is 396 RPM, and the tail rotor is 2,500 RPM.

The helicopter will lose RPMs and settle to the ground, requiring a collective pitch pull to soften the landing. If the pilot did not move the cyclic control stick the helicopter would drift to the left because of the rapid loss of translating tendency. That is because of the 1¼° left mast tilt. The helicopter will also start yawing to the right because the tail rotor RPM would decrease in proportion to the main rotor RPM.

To make a perfect hovering autorotation, the pilot should do 3 things simultaneously. Chop the throttle to flight idle, move the cyclic control stick about 1" to the right, push in the left pedal, and then pull collective pitch to soften the landing. The helicopter will then settle to the ground straight down, with no yaw, and no slide to the left. All Bell helicopters would react much the same way.

Single-rotor helicopters with clockwise rotating rotor blades would react the opposite for the cyclic movement and the tail rotor pedal controls.

Quick stop or practice autorotation.

During a power-on quick stop, the collective pitch pull will cause the downwash from the rotor to push the horizontal stabilizer down, causing a pitch up of the nose of the helicopter. (The opposite of an autorotation). The quick-stop inertia will also generate increased rotor RPM.

The pilot will want to keep this blade inertia and pull collective to maintain 100% rotor and rapidly slow the forward airspeed. With power on, he will maintain his altitude and bring the helicopter to a quick stop with power on at the desired hover height.

To properly simulate a practice autorotation, the pilot must push forward cyclic when pulling collective pitch to simulate the pitch down of the nose of the helicopter during a real autorotation, which is opposite to a quick stop.

Simulated autorotations can, therefore, cause more harm than good when the pilot is subjected to the real thing, which a pilot would get by getting factory training. I have been trained by company pilots, and they NEVER do full-on autorotations. A mishap is just too expensive. I know, I was training some police pilots in Florida with their machine, and we had a mishap. The transmission rocked too much on touchdown and damaged the centering pin on the bottom of the transmission of a Bell 206B. I was the instructor; therefore, it was my fault, and Bell paid for the part. A new transmission lowercase.

Factory training is expensive, but an accident is far more expensive than training a dozen pilots plus the added bonus is that insurance companies normally give a reduced rate to factory-trained pilots. That reduced rate is usually close to the cost of the training and expenses. With large companies that self-insure they normally have a dedicated flight instructor that is factory trained and certified annually. Liability is usually not self-insured, only the machine. No company could exist after the financial consequence of one of their helicopter pilots crashing into some people.

Autorotations cannot be taught on paper, just like a hover cannot be explained to the point where a novice can get into a helicopter and successfully hover without several hours of serious instruction.

No person had ever been born with a hover button preinstalled in his or her hands.

I know. I took a large piece of Texas to learn how to hover, and I was the second in my class to go solo.

Mast Bumping:

Mast Bumping is possible with a Bell helicopter-designed under-slung rotor system.

The models that are affected are all 2 bladed helicopters designed & built by Bell Helicopter. They are B-47 series, B-206 series, B-204, B-205, B-212, & all the B-209 Cobra or AH-1G, 1J, 1S series up to the AH-1W. The AH-1Z is not included because it was built with 4 rotor blades.

What is an under-slung rotor? The rotor hub is attached to the mast with the mast nut, or what is known in the helicopter world as the "Jesus Nut", and has a flapping axis below the point of attachment.

The motions of the rotor are rotation, pitch change (the leading edge of the rotor blades increases or decreases its angle of bite into the air), and lead/lag, which is a manual adjustment assuring that the tips of the rotor blades are perfectly aligned with the center of the mast. And flapping which is the up/down movement of the rotor blades. It is the flapping that can cause mast bumping.

Mast bumping can occur with extreme flapping of the rotor blades in the lateral plane.

Basically, not fore & aft, but left & right of the fuselage

The two ways to get into mast bumping are: (G = gravity)

1)   Attempted zero or very low "G" flight maneuvers at low airspeed.

2: Rapid lateral cyclic movements when at flat or low pitch angles with a rapid cyclic reversal.

Bell 2 bladed systems are limited in the flight manual to ½ G. That is a positive number.

What is Gravity (G)? Astronauts orbit the earth at 17,000 miles an hour at 0 G, weightless.

Alan Shepard, part of the Apollo 14 moon mission, had a collapsible Wilson 6 iron golf club and 2 golf balls with him when he went to the moon. They sailed through an airless, non-resistant vacuum with 1/6 of the gravity on earth. They went a long way for a 6 iron and the restricted movement permitted in a space suit. The 2 golf balls are still on the moon.

An underslung rotor at zero gravity can be dangerous if the pilot makes excessive, opposing, lateral movements.

1) The first way to enter mast bumping could start with a rapid cyclic climb requiring an abrupt aft cyclic movement, and then when the helicopter is pointing its nose at a high angle towards the sky, another quick forward cyclic movement that would cause an "over the top" or low "G" or zero "G" condition. This puts the helicopter in a low or zero "G" condition, and at this point, when the thrust of the main rotor is lost, the tail rotor thrust becomes the stronger power. The tail rotor thrust takes over and the helicopter starts to yaw to the right and roll right. The nose of the helicopter starts to turn right, and the top of the helicopter starts rolling to the right. The right roll occurs because the tail rotor thrust is below the rotational plane of the main rotor blades and above the helicopter's

fore/aft center of gravity. This pushes the lower part of the helicopter fuselage up and to the left.

The natural pilot reaction is to apply the left cyclic to roll the helicopter back to a level attitude. Oops, wrong movement!

The rotor system, after this action, would start rolling back to the left, and with the fuselage still rolling to the right, the main rotor mast could contact the hub stop on the main rotor hub, which is bolted to the main rotor blades. The main rotor mast of a Bell Helicopter rotor is the weakest point of the rotating mass above the transmission in a zero "G" situation.

The mast will snap.

This was the point of a fairly extensive flight test program by Bell Helicopter using a Bell 205 at their flight research center in Arlington, Texas. Dwayne Williams was the Experimental Test Pilot who flew the test. He wore a parachute, had a quick-release pilot's door, and a chase helicopter beside him at all times. Wired mast cushions were installed on the mast to record the force and angle of any impact of the hub to the mast. He was also paid a very large bonus for doing the test.

The reason behind the test was that a few AH-1 Cobra helicopters were victims of mast bumping, and the widows and families of the pilots sued Bell Helicopter.

The correct pilot movement is an aft cyclic application and if able, which should be the case, an increase in collective pitch to load the rotor.

I have flown the AH-1G Cobra & the helicopter does give sufficient warning to the subsequent problem. One must just be aware & react to what the helicopter is telling the pilot. It is not a secret; it is air sense.

When the rotor is loaded the pilot can now continue the turn to the right and add the left pedal to slow and stop the right yaw. When the yaw is slowed and the roll is arrested the pilot can now safely roll the helicopter to a level attitude, providing that he has maintained positive thrust on the rotor.

The rapidity of the roll will depend on how low an airspeed was attained, the rate of the pushover, and the amount of left pedal that was applied.

All Bell Helicopters with an underslung rotor have a maximum and minimum "G" limit. All Bell 206B and Bell 206L series helicopters have a minimum limit of +1/2 degree "G" limit. Read it correctly! The minimum "G" (Gravity) limit is a positive number. Plus, one-half of one gravity. This positive number is clearly noted in the limitations section of all these model helicopters. Less than ½

"G," zero "G," and negative "G" maneuvers are prohibited.

The mast and the main rotor hub, when at 90 degrees to each other, have 7 degrees on the left and 8 ¼ degrees on the right between the mast and the hub stop. The normal main rotor flapping in flight is 2 ¾ degrees on each side of the center when the rotor is loaded and producing thrust under normal flight conditions. The transmission and mast assembly are installed with a 5° forward tilt and a 1 ¼° left tilt. The 5° forward mast tilt permits the Bell 206 series helicopters to fly up to a maximum speed of 130 knots. This 5° forward tilt also results in a flight manual limitation of a maximum rearward airspeed of 30 knots.

The 1 ¼° left mast tilt compensates for the translating tendency of the helicopter to move to the right because of the thrust from the tail rotor. Remember, mast bumping, zero "G" maneuvers, and the tail rotor without mast thrust will push the helicopter to the right.

Translating tendency is the result of the power of the tail rotor that pushes the air to the left; the tendency is, therefore, the movement of the helicopter to the right. The 1¼° left mast tilt uses the rotor thrust to compensate for the tail rotor thrust. The mast tilt also reduces the amount of power, or the number of degrees of pitch, the tail rotor requires to hold the helicopter in a steady state hover. On a

Bell Helicopter 206 series helicopter, the difference in power between a steady-state hover and 60 knots is 8% torque on the torque gauge.

The example is easy. Find your hover torque, push forward cyclic, pass through translational lift, and accelerate to 60 knots. Look at the torque gauge; it will read 8% less torque, which is below the maximum continuous limit of 85% or 90%, depending on the model of Bell Helicopter 206 that a pilot is flying.

The mast tilt limit and translating tendency also determine the maximum slope landing capabilities of the Bell Helicopter 206.

These limits are 7 degrees left and 8 degrees to the right for slope landings. The secret is simple: slowly follow the decrease in collective pitch with a corresponding opposite move to the slope with a lateral cyclic application. If you have to land on a slope, land right high, the side of the fuel tank lid on a Bell 206 series helicopter. The helicopter has 8 degrees to work with versus 7 degrees to the left.

2) The second scenario is a high roll rate with a low-pitch setting.

The mast bumping possibility may occur if the collective pitch is bottomed rapidly with an immediate hard lateral cyclic movement. This alone would not cause mast bumping, but if the cyclic was to be rapidly reversed in the opposite direction, the inertia of the helicopter fuselage moving in one direction would not be able to react fast enough to an opposite direction input to prevent the hub stop from contacting the mast. The mast would snap, causing a vicious instantaneous lateral reaction that would snap the neck of every person in the helicopter.

The same is true if the rotor blades hit a thick wire or something solid. The blade would stop but the inertia is still there. Every person in the helicopter would die of a broken neck. In fact, I have seen the results of a rotor blade that hit a thick steel wire. That caused 4 broken necks. I went to the funeral. I knew the pilot. He was the pilot I had worked 7 on, 7 off with in the police helicopter. RIP George.

Having said that, the Bell 206 helicopter is the safest heavier-than-air flying machine, and yes that includes airplanes. Number of hours flown by all Bell 206 helicopters versus the number of deaths is the lowest in the aviation world. There may be several accidents per year, but very few deaths.

An airplane, when it crashes, is always going at least 60 to 100 knots at its slowest speed. A helicopter, if it crashes is most often at zero forward airspeed. It's not the fall that hurts; it's the sudden stop.

Long Line:

The pilot attaches a line with a cargo hook at the end to the cargo hook under the helicopter, plus an electrical plug must be plugged in to control the opening of the long line hook.

The normal industry long line is 100 feet. In an area of tall trees, this could be 200 or 300 feet. The left seat is usually the one used with a bubble window replacing the normally flat window of the helicopter's left door on medium and large helicopters.

Logging, cedar shake recovery, mining, surveying, plus many others use a long line. Forest fire bucketing is probably the most widely used application of long-line use.

The Bell 204, 205, and 212 series helicopters are the most widely used and popular machines for fire suppression. They can carry a bucket (Bambi Bucket) with an adjustable capacity of 250 imperial gallons or 2,500 pounds.

There are buckets designed for every model of the helicopter, including the much larger tandem rotor and restricted use (non-passenger carrying) Russian helicopters.

The medium Bell machines are very popular for summer long-term contracts because they can carry a 4-person fire crew, all the hose, pumps, gas, long line with a bucket, plus the personal gear and food for the crew for 2 work days at the fire. These are the first responder crews that are contracted for a 10-minute alert response.

Every day or immediately after the passage of a thunderstorm with lightning, a fire patrol fixed wing patrol aircraft is sent out to locate any fires.

Retreating blade stall:

Retreating blade stall is because of too much airspeed in a helicopter, basically past 1.1 of the VNE. Velocity Never Exceed in a Cobra helicopter was 190 knots. Add .1 or 19 knots gives 209 knots. (The Bell AH-1G Cobra was designated the Bell 209. Curious!) Exceed that speed, and the retreating blade will start a high-speed stall.

The AH-1G Huey Cobra was a victim of retreating blade stall when in use in Vietnam. A gun run was normally where this was encountered.

The rotor diameter of the AH-1G Cobra is 44 feet. (13.4 m). The VNE (Velocity Never Exceed) was 190 knots. With the Cobra in a dive, the VNE airspeed could be exceeded using the gun run torque of 33 lbs (Max 50 lbs). When this happened, the rotor would enter a condition of retreating blade stall. The pitch angle of the rotor would be too high on the left side, the retreating blade causing a stall. The right side of the rotor would be the most efficient, producing the most lift. Normally, that should result in a left roll. The difference was the speed of rotation of the blades. The higher the speed of rotation, the greater the angle of precession. The physics would enter the equation. The greater lift force is on the right; the torque is rotating the blade to the left in front of the pilot. Precession would then cause the force to act almost $90°$ to the left of the lifting force because of the rotational speed of the rotor, which was 324 RPM. The force would then be mostly back, causing the nose to lift, reducing the airspeed. The pilot would need to reduce power to decrease airspeed, raise the nose to lower airspeed, and exit the dive, and the retreating blade stall. The results of the precession would initiate the rising of the nose for the pilot.

High speed will cause the blade tips to glow with heat. The shock wave on the advancing blade heats the blade tips enough that, using NVGs (Night Vision Goggles), the glowing blade tips are readily apparent. The shock wave also causes blade slap, which is why some helicopters can be heard from a long way away, especially in clear, cold air. In the Arctic, the Bell 214ST heavy blade slap can be heard over 20 miles ahead of the helicopter.

In Vietnam, the blade slap from a Bell UH-I Huey could also be heard a long way away and was often described as "the sound of freedom."

The maximum speed of a helicopter is determined by the speed of sound.

The rotor speed of the UH-I helicopter is 324 rpm with a 48-foot diameter rotor and a VNE of 148 mph or 130 knots, and the AH-1G Huey Cobra has a rotor speed of 324 rpm with a 44-foot diameter rotor with a VNE of 190 knots.

The loss of 4 feet of the diameter of the Cobra rotor increases the VNE by 60 knots.

The VNE of and helicopter is determined by experimental flight testing during the development stage of that particular helicopter. Aerodynamics during the design stage tells the engineers the VNE before the machine is even built.

That is then tested during test flights, and the maximum speed of the rotor is determined. The true VNE is 1.1 of the placarded VNE in the flight manual.

That means the Cobra could actually go 19 knots faster before entering a retreating blade stall. I know, I have over 450 hours in the Cobra.

I have several thousand hours in a Bell 407, and it easily cruises at 140 knots; it's VNE, which means that above 1.1, at 154 knots, the blades would also enter retreating blade stall, and they do. That's what test pilots do, they test stuff.

LTE: Loss of tail rotor effectiveness:

Bell helicopters can get into loss of tail rotor effectiveness in 3 ways.

The first is attempting a sideslip like an airplane. Lower the nose of the helicopter, lower the collective power, and push in the right pedal.

The tail rotor is on the left-hand side of the vertical fin, effectively pushing the uprising air down and away from it. This creates turbulence around the tail rotor and the vertical fin. No real control problem exists, but the descent will be irregular and hard to control in the yaw axis.

To sideslip a Bell Helicopter, lower the nose, lower the collective, and push in the left pedal. The power of the tail rotor thrust is now up, augmenting the rising air through the tail rotor. This is a very stable manoeuvre and the helicopter will descend rapidly. To terminate the sideslip the pilot needs only to raise the nose of the helicopter to a level attitude while simultaneously pushing in the right pedal to bring the nose back to an into wind condition and a coordinated, ball centered, on the Turn and Slip Indicator. A slight rise in collective and the helicopter will resume a normal descent speed and attitude.

The second way is to turn the helicopter about 30 degrees to the right with the tail into wind. This puts the tail rotor under the descending main rotor vortices, causing an erratic control problem.

The third way is to turn the tail into wind causing the wind to erratically shift from side to side of the tail rotor, again resulting in an erratic tail rotor control problem.

This LTE problem can manifest itself in a hover, in a downwind descent in flight, and in a right-hand pedal turn in a tailwind hover in a high wind condition. The effect in a left-hand pedal turn under the same circumstances is greatly lessened because the tail rotor is producing a much higher power or thrust level that counteracts the LTE effects, and the tail rotor is not under the blade tip vortices long enough to create a control problem.

Engine Lag:

One of the problems with a turbine-powered helicopter is engine lag.

A two-stroke piston engine will give power instantly, whereas a turbine engine has to spool up to give power.

A piston engine normally operates between 2,000 to 10,000 or more RPM.

A turbine engine can operate up to 30,000 to 60,000 RPM. Several dozens of pounds of compressor and turbine stages take a few seconds to increase to 10,000 or more RPM. Turbines are built as light as possible for that reason, plus the resistance to high heat.

Engine lag is the amount of time it takes from the moment the power requirement is demanded until the engine can give the power to the rotor system.

Find a large open area with no obstructions and little or no wind.

Bring the helicopter to a stable hover, with a machine that is lightly loaded and stabilized at a 5 to 6-foot hover. Note that your torque is below maximum continuous power, and especially note your arm position.

Lower the helicopter to the ground, very light on the skids, so that you are actually still flying, but just brushing the tops of the grass.

With your feet firmly placed on the tail rotor pedals, make a conscious, willed effort not to move the pedals until after the helicopter has yawed through 30 to 45 degrees before moving the power pedal to counteract the power input.

Abruptly pull in collective pitch until you are just below the previously noted arm position.

Hold that power until the yaw starts while counting the time it takes before the helicopter yaws 30 to 45 degrees, enough to require the power pedal input. The time will be 1 ½ to 2 seconds.

The time is called engine lag and that time is critical when operating a turbine-powered helicopter.

When test flying, I would often make my approach to the runway, and upon pulling power to hover I would deliberately not push in the left pedal, letting the torque turn the helicopter to the right. Then when pointing towards my proposed landing spot I would then push in the left pedal to stop the turn. This required less power to hover. The first power application would be with the collective and the second application would be the small power requirement from the left pedal. This made for a much smoother approach and turn.

This way the pilot can leave translational lift in a lower power co-ordination of the two moves.

Downwind:

Downwind approaches are not a good idea; however, there are occasional circumstances when that type of approach cannot be avoided. If that is the case, then do it right, do it safely.

Referring to the turn and torque part, I will use many of the points of that writing to discuss the downwind approach.

I have demonstrated to many other pilots the dangers of a downwind approach terminating in a right-hand turn to a hover.

In a downwind approach, I will always try to bring back the airspeed to as slow as possible to bring the helicopter into a high hover long before getting close to the ground where a turn to a hover is done using the power pedal, the left one in American made helicopters. This technique will give the pilot the option of flying through the approach and getting out safely or coming in very slowly with high power all the way to the ground to prevent LTE, loss of tail rotor effectiveness, or settling with power.

The only way out of settling with power is to lower the nose of the helicopter, lower the collective, and fly out of the phenomenon until translational lift is regained, permitting an increase in power to fly out of the situation.

The best, safest way to do it in a helicopter with the blades with counterclockwise rotation is to come in from the downwind approach and terminate with a left turn. The power is much more controlled.

Initiate the approach = lower power/torque.

With decreasing speed, the pilot adds power to the left pedal and adds torque.

Descending using less torque.

Losing translational lift, the pilot adds collective power/torque.

Engine lag happens, slowing the need for more power.

Then left pedal is required to enter a left turn, and more power/torque is added.

The rotor then settles to an into wind hover situation where the need for the left pedal decreases, less power/torque.

Then hover power is required, more torque.

Note the power sequence. Less power, more power, less power, more power, less power, until the hover power is required.

From downwind to a hover with a right pedal turn.

The sequence is:

Lower collective, less power/torque,

Start the right-hand turn, less power/torque.

Lower the collective to enter the hover less power/torque.

Now, you need it all at once.

Stop the descent=more power/torque.

Left pedal to stop the right-hand rotation=power/more torque.

Lose translational lift=more power/torque.

More power= more left pedal= more power/torque.

Usually, at this point, the power/torque requirement would exceed the maximum power/torque limit of the transmission. The throttle must be closed to idle to prevent an over-torque, forcing the pilot to finish the approach with a hovering autorotation. Hopefully, the surface below the helicopter would permit that.

The one other thing that is important is that with a left pedal turn to the hover is your tail rotor. The tail rotor will be turning into an area that you can clearly see in your front and to your right. And in a Bell helicopter, the vertical fin is on the right side, so if you did hit something, it would be the vertical fin on the right rear side of the tail and not the tail rotor, which is on the left side of the vertical fin.

That changes with later models 212 and 412's that have a tractor tail rotor. The tail rotor is on the right side of the tail fin.

NEODD SWEVEN:

A red hawk at 7,000 ft when I was eastbound to Baie Comeau took me by surprise.

I was tasked to go to Baie Comeau from Québec City to look for a person who was lost in the bush. I checked the weather & found that a wind reversal took place at 5,000 feet. Below 5,000 feet, there was a strong wind from the East, a headwind. Above 5,000 feet the wind reversed to westbound at over 20 knots, a tailwind. (23 mph). So, going by the aviation rule of NEODD SWEVEN, I climbed to 7,000 feet.

NEODD = North to East, meaning from 000 degrees magnetic to 179 degrees magnetic, I would fly at an odd-numbered altitude. The wind shear was reported at 5,000 feet. At that altitude, there would be a lot of turbulence.

SWEVEN = south to west, even altitudes. 180 degrees to 359 degrees.

I climbed to 7,000 feet & got a 20-knot push instead of a 15 to 20-knot headwind. A pilot no-brainer when you are doing only 120 knots. A 40+

knot difference in ground speed and that is a lot for a helicopter.

An object that is on a direct collision course has no relative movement. Only color or size can be detected.

A red hawk weighs about 5 or 6 pounds. It is a raptor that is found throughout North America. I do not know if it was beak to nose with me or if I came up behind it. I saw it only a few feet in front of the helicopter. At 7,000 feet, it was a very rude and dangerous surprise. I jerked my cyclic control stick to the left, which meant that I tried to put the bird on my right side. Reflex, automatic reaction. If I had rolled right, the hawk would have, without a doubt, gone through my co-pilot windshield. If I had not rolled to the left, the hawk would have come through the pilot-side windshield at about chest height. The hawk blew past so close that if I had stuck my hand up to my wrist, out the pilot side sliding window, it would have taken off my fingers. 5 to 6 lbs at 120 knots is a lot of force. 7,000 feet is a long way to fall if you are all cut up, out of control, or unconscious.

Turn & Torque:

I will use a Bell Helicopter for this example for I believe that most helicopter pilots have flown a Bell product sometime in their career.

A helicopter in a steady hover is the pilot balancing all the forces affecting the machine to keep it over the same place.

There are several: the wind, the power from the engine that turns the main rotor and creates torque, the tail rotor counteracting the torque from the main rotor, mast tilt, and the constantly changing cushion of air under the helicopter when in ground effect. This hover is the time and place when the pilot has to work the hardest to keep everything balanced.

In a steady into wind hover, if the pilot wants to increase the height above ground, he must increase collective pitch. This increases torque, so the helicopter would have a tendency to yaw to the right. To counter this reaction, the pilot must increase the left pedal, or to better understand the function, I could call it the power pedal. Demand power from the collective & that would require adding power to the tail rotor for more power to counteract the increase in torque by the main rotor blades and still maintain your same position over the earth.

Should the pilot desire to turn the nose to the left or to the right, several other forces come into play.

The first thing to note is that the pilot sits on the right-hand side of the front two seats in a Bell 206. The pilot's visibility to the right rear is excellent, any obstructions to the turn would be visible. On the other hand, the machine itself obscures the left rear side of the helicopter.

The tail rotor is on the left side of the tail of the machine. A left pedal turn would give the pilot excellent visibility over what the tail will pass, with the added bonus of the tail rotor itself being on the left side. If something did hit the rear, it would most likely hit the vertical fin first.

A turn to the left, increasing power pedal input will cause a small momentary increase in torque on the torque gauge and a slight momentary decrease in height above the ground. The amount will depend on how fast the pilot initiated the pedal input, and how fast he wants to turn. After that, the faster the turn, the more that the helicopter will increase in height and the pilot will need to lower collective to maintain height above ground. The reason is simple: the Bell rotor turns counterclockwise, passing above the pilot from right to left. A left turn increases rotor speed by a few RPM, causing the helicopter to climb. Torque increased slightly when the turn was initiated up to about 90 degrees, then because the increase in rotor RPM caused the helicopter to climb, the pilot would slightly lower the collective pitch to maintain height above ground, which by this time, the helicopter would be about 180 degrees into the turn. The helicopter would now be directly tail into wind, which may cause the helicopter to dip its nose down slightly, which would cause an aft cyclic movement to stay over the same place, plus a small increase in power. By now, the helicopter would be about 270 degrees into the turn, and the wind would now be pushing on the vertical fin, helping the helicopter in the turn; again a slight decrease in power would be required. The helicopter would now have finished the turn, and the pilot would need to push in the right pedal to stop the turn, but because the turn has stopped, the increase in RPM is now gone, so the pilot would need to add power to retain the hover height when steady state into wind.

During the entire turn to the left, no two power applications were made without a decrease in power between them.

A right-hand turn puts the tail rotor itself into the path of any obstruction to the turn.

The right turn starts with a right pedal application, a decrease in torque, and the decreased rotor RPM by about 90 degrees will cause the helicopter to settle, requiring an increase in collective pitch or torque. At 180 degrees of turn, the tail into wind requires another increase in torque, which increases the speed of the turn; the pilot must now apply more power to the tail rotor to slow the turn. At 270 degrees into the right-hand turn, the wind will start pushing on the vertical fin, increasing the speed of the turn, or RPM, and another collective input, more power is needed. This will require a power pedal input, just as the helicopter is coming out of the turn and back into wind. Stopping the turn will require more power pedal input. Stopping the turn during the last few degrees of the turn required three power applications in a row. For a heavy machine, this could result in an over-torque. A right-hand turn is a succession of several reductions of power in a row to several increases of power in a row.

Given the option, a pilot should always try to make any 180-degree turns to the left, safety and power management are the reasons.

Power management is self-evident, but why safety?

A helicopter in a left-hand turn will have the tail rotor move over what the pilot has looked at, whereas in a right-hand turn for the first 90 degrees, the pilot may not have seen an obstruction. If the tail rotor should strike an object, post wire, etc., the helicopter would continue the right yaw, and it will accelerate. The increase in yaw rate would be rapidly noticed by the pilot, but the amount of time required for that to happen and then for the pilot to react to the increased yaw rate would consume precious seconds that would only result in the helicopter attaining a dangerously rapid spin that would require a quick removal of power into a hovering autorotation with no way to physically stop the yaw rate. A rollover and crash would be the possible result if the right yaw was not stopped.

The helicopter in a left-hand turn will give the pilot a much better chance to react and control the yaw rate. A tail rotor strike or failure would result in the helicopter stopping the left yaw, and then starting a rapid right yaw. Pushing the left pedal and turning right would result in an immediate reaction by the pilot before the yaw rate has had time to get to a high rate of degrees per second of the right yaw.

This would be true with all helicopters with a counterclockwise turning rotor when seen from above.

PART 3: SPEED

What is a simple definition of speed?

Speed is defined as: **The rate of change of position of an object in any direction**. Speed is measured as the ratio of distance to the time in which the distance was traveled. Speed is a scalar (like a ruler) quantity as it has only direction and no magnitude. (size)

Meters per second equals meter/s, or 1 kilometer per second (kps) = 1000 mps.

Miles per second equals mps. Miles per hour equals mph.

1 mile = 5,280 feet or 1760 yards

I nautical mile is 6073 feet 1.386 inches

I kilometer is 1000 meters or 3280 feet 10.079 inches

1 mile is 1.609 kilometers

I kilometer is 0.621 miles

I inch is 2.54 centimeters

1 centimeter is 0.3937 inches

Speed of light: (c). The fastest object in our known universe is 299,792,458 meters per second. 186,282,397 miles per second or 670,616,629 mph.

Light is energy. Energy is light. Light consists of photons and does not have mass.

MIT has recently developed a camera that can take 1 trillion frames per second. It can photograph the light (protons) as they move. (femto-photography)

One light year is 5.88 trillion miles or 9,460,730,000,000 km.

Or solar system is 2.5 million light years from Andromeda, the nearest galaxy.

(In other words, the closest alien from outside our solar system would have to travel 2.5 million years at the speed of light to reach our planet.) ET would be a really old man.

Energy is the only thing that can travel at the speed of light

Light can travel around planet Earth 7 ½ times in 1 second

Light from our sun takes 8 1/3 minutes to reach Earth, at 93 million miles away, or 149,668,992 million kilometers.

1AU = distance from sun to earth. 1 Astronautical Unit = 93 million miles.

Light from the sun to Pluto 5 hours 10 minutes.

Light from our sun to the nearest star (Alpha Centauri) 4.24 light-years

The sun and our solar system rotate in the Milky Way galaxy.

Fastest things in the universe.

1) Expansion of the universe: the only thing faster than the speed of light because it has no mass.

2) Light

3) Gravitational waves

4) Cosmic rays

5) Blazar jets

6) Dark

Light cannot exist without darkness.

The universe is expanding at 3.26 million light years per second and is constantly increasing its expansion faster than the speed of light. Einstein's cosmic speed limit only refers to the motion of a physical object through that space, including light.

Blazar jets can emit extremely high-energy gamma rays, as seen by the Chandra X-ray Observatory.

Speed of Dark: Dark is the absence of light. The moment that light leaves, darkness returns. Therefore, darkness has the same speed as light.

Darkness rises up from the earth. Up buildings or mountains. The bottoms of clouds are often pink with sunlight when the earth is dark. Why? Earth is almost round.

Energy equals mass times the square of the speed of light. $E=mc^2$, Albert Einstein, 1905.

For a mass to obtain the speed of light, it must transform itself into energy.

When an atomic bomb detonates, the plutonium transforms its mass into energy.

Shake = 10 nanoseconds.

1 nanosecond is 1 billionth of a second. (This term was made in 1940 during the Manhattan Project when building the atomic bomb.)

Light will travel almost 1 foot or 30 cm in 1 nanosecond.

1 Shake equals 10 nanoseconds, = 9 feet or 3 meters.

3 shakes. The atomic bomb starts and ends in detonation. (Nuclear fission)

I'll say that again: a nuclear explosion lasts 3 shakes, which is 30 nanoseconds, which is 27 feet. Everything that happens, the flash, the noise, the explosive displacement of everything near it, the radiation, are all results of the 3 shakes. The nuclear explosion is complete after 3 shakes, no matter what the size. The nuclear explosion is so violent because it happens so fast.

$E=mc^2$. E=energy. m = mass. c = speed of light. 2 = speed of light squared.

E.g: Energy = mass (10 lbs), X c, 186,000 X 186,000. The E, =Energy is 3 shakes, a nuclear explosion. 10 x 186,000 x 186,000 = the energy produced by fission, the splitting of the atoms.

Our sun, like any star, is powered by nuclear fusion, atoms fusing together, the opposite of fission, and is not yet available for use commercially.

$E=mc^2$. Energy = mass times the speed of light squared.

During the Second World War, the United States dropped 2 atomic bombs, one each on Hiroshima and Nagasaki, Japan. They were so devastating they ended the war.

No other atomic bomb has been dropped in anger. Planetwide nuclear winter could result in escalation or retaliation.

Global thermonuclear war is like Tic Tac Toe. Unwinnable.

Little Boy, the nuclear bomb dropped on Hiroshima, was a gun-assembly fission bomb using uranium U-235. Little Boy weighed about 9,000 lbs and produced an explosive force of 15,000 tons of TNT, in 3 shakes.

Fat Man, the nuclear bomb dropped on Nagasaki, was an implosion fission bomb using plutonium Pu239. Fat Man produced 25,000 tons of force in 3 shakes.

The most efficient nuclear bomb today is the B41 using Uranium U-238.

B41 weighed 4800 kg or 10,582 lbs and yielded 25 Mt. (megaton = 1 million tons of force of TNT.) [TNT = Trinitrotoluene = $C_7H_5N_3O_6$]

The distance to escape Earth's gravity is 13,000,000 miles or 21,000,000 km.

An object will never escape the Sun's gravitational force. The Sun's gravity extends as far as light has had time to travel since the Sun came into existence 4.5 billion light years ago.

Lightning is electricity, light, energy. Electricity is measured by humans in Volts (electrical pressure or amount), Ampere (amp, current, or rate of flow), and Ohms (electrical resistance through which it flows). The best electrical conductors, in order, are silver, copper, gold, brass, aluminum, zinc, and nickel. Of the five best electrical conductors, gold is preferred because it never tarnishes. All others must be sheathed to prevent tarnishing or corrosion and that adds weight and complication to the building of any circuits. Circuit boards are always made with gold.

Turning on a switch to get electricity to power a light, TV, or other appliance requires a huge infrastructure from a power plant or dam, transmission lines, and distribution points to the transformer and wires to your home. Then you can use the electricity generated by the energy of falling water to create light. The energy produced is used in the same microsecond that it is produced because the speed of electricity, light, and energy is 186,000 mph or 300,000 kph per second. (299,792,458 meters per second) (remember, light can circle the earth 7 ½ times in one second)

Lightning is caused by the positive and negative charges between the cloud and another cloud or the ground. A giant spark of energy at the speed of light is produced. When the resistance of the air breaks down between the positive and negative charges, there is a rapid discharge of electricity between the cloud and the ground, or cloud to cloud.

This giant spark of energy produces light. This energy ignites the air. The air explodes, causing thunder. The noise of the thunder travels at the speed of sound or one mile in 4.69 seconds. So, at approximately one mile per second, if you see lightning, start counting the seconds. If you can count to 5, that means that the sound of the lightning bolt you saw travelled through the air about 1 mile away.

After the first lightning bolt, you could, by counting the seconds, always know how far away the lightning is hitting the ground or another cloud.

To get kilometers, multiply each second times 1.6.

Lightning strikes the earth on average 8.6 million times per day.

Lake Maracaibo in Venezuela is the lightning strike capital of the world. 140 to 160 nights per year, an average of 28 lightning strikes per minute for up to 10 hours at a time.

SLOWER THAN LIGHT

The fastest man-made rocket was the US NASA Parker Solar Probe as it approached the sun at 394,736 mph or 635,266 kph on 18 Oct 2023.

Fastest speed by a human. Apollo 10 on its way to the moon at 24,791 mph or 39,897 kph.

Geostationary orbit or Geosynchronous Equatorial Orbit (GEO) is 22,236 miles or 35,786 km.

International space station circles the earth every 90 minutes at 17,500 mph or 28,000 kph at between 200 and 250 miles above Earth's surface.

Any living, breathing animal (humans) must wear a pressurized space suit or be in a pressurized vehicle in space or near space. Blood pressure by the heart would explode the body at zero pressure. Earth's air pressure is 14.7 pounds per square inch at sea level and decreases with altitude.

Intercontinental Ballistic Missile (ICBM) 15,000 mph or 24,000 kph.

Hypersonic is 5 times the speed of sound.

Speed of sound in air. The denser the medium, the faster the sound moves.

767 mph / 343 mps at sea level; 68 deg F or 20 deg C. 1100 ft/sec.

Speed of sound in salt water;

4760 to 5150 ft/sec or 1450 to 1570 mps.

Speed increases with temperature, salinity, and pressure.

The speed of sound through steel is 17 times of air.

13,332 mph and 5960 mps.

Fastest air-breathing aircraft. SR-71 at over Mach 3.3+ and 90,000+ ft. It had over 4000 missiles shot at it. It outran everyone shot at it, no SR-71 was ever shot down. The Mig 25, made of stainless steel, was specifically designed by the Russians to try to shoot down the SR-71. It failed. The SR-71 fuselage expanded over 4 inches (11cm) from nose to tail because of the heat produced by its speed. SR-71 is 107 ft, 6 in long.

SR-71 maximum speed was determined by the fuselage heat, not power.

SR-71 missions were long, and the pilots used to heat their lunches by putting them close to the windshield.

Low-Earth orbit satellites and high-altitude drones have replaced the SR-71 and U2.

The F-117A was the first stealth aircraft.   One was shot down in Serbia.

A B2 bomber was never shot down. Only 20 were built at a cost of over $1 Billion each.

Speed of Commercial aircraft. Speed of sound 767 mph, 1235 kph, 667 knots, 1,125 ft/sec, 343 m/s, or 1 km in 2.91 sec or 1 mile in 4.69 sec.

Airbus 380 - 670 mph or 1,000 kph.

Taxi @ 30 -35 kph. 18 -22 mph

Take off @ 240 – 280 kph. 149 – 177 mph.

Cruise @ 830kph, 515 mph.

A350 and B747 @ 900 kph or 550 mph at 31,000 to 38,000 feet.

Lands at 270 – 290 kph. 160 -180 mph.

Average speed and altitude.

A320 @ 967 kph, 601 mph at 36,000 ft.

A330 @ 950 kph, 590 mph at 36,000 ft.

A380 @ 1041 kph, 647 mph at 43,100 ft.

Boeing 747 @ 920 kph, 571 mph at 35,100 ft.

B737 @ 850 kph, 528 mph at 35,000 ft.

B777 @ 905 kph, 562 mph at 35,000 ft.

B787 @ 907 kph, 563 mph at 40,000 ft.

Embraer 190 @ 829 kph, 515 mph at 39,370 ft

E175 @ 797 kph, 495 mph at 39,370 ft.

The Earth rotates once a day at 1000 mph or 1600 kph at the equator.

The earth is 24,901 miles (40,075 km) in circumference at the equator. Plus, it is 7,926 miles or 12,756 km in diameter at the equator and 7,907 (12.725 km) in diameter from the north pole to the south pole.

The north pole to the south pole is 12,436 mi or 20,014 km following the surface.

Earth's orbital speed around the sun is 67,000 mph or 107,000 kph, and takes 365 days to make one revolution around the sun. Once every 4 years, one day is added because of a small variation and is called a leap year.

The Julian or Christian calendar was invented by Julius Caesar around the time of Christ.

There are 8760 hours in 1 year, add 24 hours for a leap year.

If a pilot has over 17,000 hours of flying time, he has over 2 years in the air.

I have over 15,000 hours helicopters.

The most fun I ever had in the air was in Pitts Special doing aerobatics over the Florida Everglades.

The earth is tilted 23½ ° in relation to the sun. During the spring and fall equinox, the angle is 0° at the equator. During the summer solstice, the North Pole is tilted 23½° towards the sun (Tropic of Cancer), and during the winter solstice, the South Pole is tilted 23½° towards the sun (Tropic of Capricorn)

The Tropic of Cancer was named because the sun seemed to be in the Cancer constellation 23½° North of the equator, and the Tropic of Capricorn because the sun seemed to be in the Capricorn constellation 23½° south of the equator.

Then we have the Arctic and Antarctic circles, each 23½ ° from the axis of rotation of the earth. The Arctic comes from the Greek word Arctos, or "Arcturus," meaning "guardian of the bear." This is the North Star over the North Pole, around which the Earth rotates. Polaris, or the North Star, is at the end of the handle of the Big Dipper which is part of the Ursa Major or "Great Bear" constellation, hence, Arctic. Under which are found the Great Bears, the polar bears, the largest species of bear, and land carnivores. (All polar bears are left-handed, and their skin is black). The North Pole is also where the earth's magnetic lines of force turn above the earth down to the South Pole. These lines of force divide the earth magnetically and are what is used by a compass for direction. The magnetic poles move about 34 miles or 55 km every year. I have explained "variation or deviation", the angular difference, depending on where you are on Earth, the number of degrees between the north pole and the north magnetic pole with a compass.

The earth rotates at 24,000 mph or 38,624 kph. At 1,000 mph this divides the earth into 24 hours for our clocks and 360° for our compass.

The Aurora Borealis or Northern Lights are caused by the sun sending out huge clouds of electrically charged particles which are attracted by the magnetic lines of force coming from the poles, which is why you can only see the Auroras near the north or south poles where the magnetic lines of force come together and curve down to the poles. Aurora comes from the Roman goddess of the dawn.

We also have the Aurora Australis or southern lights.

At the opposite end of the planet, we have the South Pole and the Antarctic.

(Anti Arctos) Antarctic means no bears. Antarctica is also the driest place on earth. The Atacama Desert in Chile and the McMurdo Dry Valleys in Antarctica never rain or snow.

68% of Earth's land mass is north of the equator, and 32% is south of the equator.

Oh, and don't forget Santa Clause lives at the North Pole with his elves and reindeer.

Speed of an air-driven dentist's drill averages between 25,000 to 80,000 rpm.

The drill head or burr can be up to 180,000 rpm or 3,000 revolutions a second.

Some can go up to 400,000 to 600,000 rpm for special drill jobs. Tiny holes require higher speeds because of the hardness of the tooth and the lack of torque of the drill bit.

The fastest bullet is the .220 Swift at 4,665 ft per second or 1,422 meters per second.

20 mm bullet = 3,380 ft/sec.

.50 calibre bullet is max at 3,450 ft/sec but most are 2,700 fps to 2,900 fps with a chamber pressure of 55,000 lbs or 3,792 bar.

.45 cal is 830 fps (253 mps) from an M1911A1 pistol and are subsonic. The bullets are below 1100 fps and are ideal for use with a silencer. (The bullet travels slower than the speed of sound). A pistol (revolver) cannot use a silencer because the gas pressure and noise escape from around the rotating cylinder.

World's fastest train:

China has the 3 fastest trains in the world, all maglev. (Magnetic Levitation)

The fastest is the Shanghai Maglev, with a speed of 460 kph or 285.8 mph.

World's fastest production cars:

Koenigsegg Jesco Absolut = 330 mph (Aug 2023)

Hennessy Venom F5 = 311 mph

Bugatti Bolide = 310 mph

Bugatti Chiron Super Sport 300 (305 mph)

2024 Corvette ZR1, 1,064 HP, 828 lb-ft of torque and 215 mph. $180,000 US

Fastest (not street legal) Thrust SSC 763 mph (1,227.9 kph) and the first car to break the sound barrier.

The average spin rate of a car tire is 840 rpm at 60 mph with a 2-foot diameter tire.

1,120 rpm at 80 mph and 1,400 rpm at 100 mph.

The speed rating of a tire is a letter from A to Z, ranging from 3 mph (5 kph) to 186 mph (300 kph).

Fastest car tire. Michelin Pilot Sport Cup 2.

These tires were tested on a machine specifically designed to test the space shuttle tires up to 317 mph or 510 kph. Stamped BG and made specially for the Bugatti with a cost of $42,000 for 1 set of 4 which would only last 15 minutes at 250 mph or 402 kph.

Highest wind speed on Barrow Island at 254 mph, 408 kph, wind gust during Tropical Cyclone Olivia. (50 miles NW of Australia)

The highest wind speed recorded by a man was on Mount Washington, Vermont, at 231 mph, 371.7 kph.

Highest estimated tornado speed at 318 mph, 512 kph.

Eastbound Polar jet stream can be as fast as 200 mph, 321 kph, and often over 100 mph, 160.9 kph.

Beaufort wind scale has 12 categories from 0, calm, to force 12 up to 83 mph. Above a force of 12 is the hurricane scale.

Speed of a hurricane using the Saffir-Simpson Hurricane sustained wind scale.

Category One: Winds 74 to 95 mph.

Category Two: Winds 96 to 110 mph.

Category Three: Winds 111 to 130 mph.

Category Four: Winds 131 to 155 mph.

Category Five: Winds greater than 155 mph.

Tornado wind speeds:

Fujita Scale of Tornado Intensity.

F1 - 73 to 110 mph.

F2 - 111 to 135 mph.

F3 - 136 to 165 mph.

F4 - 166 to 200 mph.

F5 - >200 mph.

Fastest military helicopter is the Westland Lynx at 249 mph or 400 kph.

Fastest US helicopter is the tandem rotor CH47F Chinook at 200 mph or 321 kph.

Fastest boat: (1 knot =1.151 mph)

Jet-powered hydroplane: Spirit of Australia. 275.9 knots, 317 mph, 510 kph.

Fastest US warship: USS Detroit. 47 knots, 54 mph, 87 kph.

USS Ronald Reagan, 1092 ft, 332 m, aircraft carrier. 31.5 knots, over 101,400 tons.

USS Gerald R Ford, 1105 ft, 337 m, 31 knots, the largest warship ever built.

Both have over 4,500+ crew. Can carry up to 90 aircraft.

Imagine the nuclear power required to move over 100,000 tons to over 30 knots.

4 screws, each 21 ft or 6.4 m in diameter at over 40 tons, 40,000 kg, each using 2 140,000 horsepower nuclear reactors. The anchor weighs 30,000 lbs with 1,440 ft of chain and each link weighs 136 pounds.

Largest cruise ship: Wonder of the Seas, 1,188 ft, 362 m long. 6,988 passengers, 2,300 crew, 18 decks, 236,857 tons. Burns 1 liter of fuel oil every 5 inches when cruising. Imagine the amount of food needed to feed over 9,000 people 3 times each day.

Largest ship: Seawise Giant, a supertanker, 1504 ft, 458 m at 260,000 tons. Two 50,000 hp engines, 16.5 knots, or 19.0 mph or 30.6 kph. Can carry 4.1 million barrels or 564,763 tons of crude oil.

Sailboat maximum speed in knots

= 1.34 X square root of the waterline length in feet.

= 2.43 X square root of the waterline length in meters.

And then there are speed bumps and potholes.

Speed bump in Norwegian is FARTSDUMPER.

Pot hole is chicken nest, "nid de poule" (knee de pool) in French.

Imagine driving on a road full of Fartsdumpers and chicken nests?

Fartsdumpers:

3 highest mountains.

Mount Everest: 29,035 ft, 8,849 m, above sea level.

Mount Mauna Kea, Hawaii: from base of mountain to the top, 33,476 ft. 11,203 m.

Mount Chimborazo, Ecuador, South America, 20,548 ft above sea level plus 7096 ft higher than Mount Everest @ 36,131 ft, or 11,012 m. Highest because Mount Chimborazo is on the Equator and the earth is a spheroid.

The top of Mount Chimborazo is 6,384.4 km or 3,967.1 mi from Earth's center.

Mount Everest is 6,382.3 km or 3,365.8 miles from the Earth's center.

Chicken nests:

Lowest place on earth is the Dead Sea at 1,358 ft, or 414 m below sea level (BSL), between Jordan and Israel. (2,388 ft or 728 m depth BSL at the bottom of the lake)

Lake Assai, Saline Lake, Djibouti is 515 ft or 157 m below sea level in Africa.

Death Valley is 282 ft or 86 m below sea level in North America.

Mariana Trench is 36,037 ft. 10,984 m deep. In Pacific Ocean, 124 mi, 200 km east of the Mariana Islands, 1,584 mi, 2,500 km long & 42 mi, 69 km wide.

James Cameron's (Author of Abyss and Avatar) craft in the Mariana Trench was subjected to 16,000 lbs per sq in of water pressure or 1,125 kilograms per sq cm. Surface pressure is 14.7 lbs per sq in.

Fastest bird: Peregrine falcon: 240 mph, (386 kph) in a dive, 86 mph in level flight.

The fastest flightless running bird is the ostrich at 40 mph (60 kph). The Roadrunner is the fastest running bird at 26 mph (42 kph) that can fly. The coyote can run faster at 56 mph (69 kph).

Next time you eat some chicken, you will know that its direct ancestor was 55 to 80 million years ago Tyrannosaurs rex or T-rex, the big speedy villain in "Jurassic Park".

Fastest animal:

Cheetah 60 to 70 mph.

Pronghorn antelope, nicknamed "speed goat," is the second fastest land animal.

Fastest serve in tennis, Sam Groth at 163.7 mph or 263.4 kph.

Average professional tennis serve is 100 to 120 mph. (193 kph)

Most pros use 98 to 100square-inch tennis rackets for their balance and speed.

Normal grip size is 4 ¼ (2) for women and 4 $\frac{3}{8}$ (3) for men.

Fastest baseball has been calculated to be 125 mph, but only 106 mph has been attained.

Fastest hockey puck was 110 mph, 177 kph by Russian Denis Kuylash

The average spin rate of a football is 650 rpm. NFL QBs average in the 700 to 800 rpm range. The faster spin rate means less resistance as the ball moves through the air and will go faster and further.

Gravities terminal velocity of a human being is 118 mph or 190 kph.

Speed of a golf ball:

Pro golfers:

Ball 160 -180 mph, or 257 to 274 kph.

Driver – 167 mph

3 iron – 142 mph

6 iron 127 mph

Wedge – 102 mph.

1 mph (1.6 kph) = 2 yards (2 meters) of distance

Average man = 147 mph or less.

Average woman = 125 mph or less.

Dimples on a golf ball vibrate the air around the golf ball, decreasing drag and permitting the airflow to hold around the ball, decreasing the size of the wake turbulence, decreasing drag, and increasing the distance the ball will fly.

Dimples determine velocity, launch, and spin rate, which increase speed, therefore, distance.

Number of dimples: Pro V1 has 388 dimples, and ProV1x has 348. TaylorMade has 322, and Carver golf balls have 332 dimples.

The ball should have an evenly spaced dimple pattern with one shallow and one deep dimple.

Longest golf balls at 115 mph.

5. Srixon Z-star Diamond at 344.8 yards.

4. Vice Pro Plus at 345 yards.

3. Mizuno RB Tour X

2. Titleist Pro V1x

1. Titleist Pro V1x Left Dash at 348 yards.

84% of the world's countries have golf courses. The are 195 countries in the world.

There are 1.2 to 2 billion golf balls lost every year in the world,

1.2 to 2 billion golf balls are manufactured each year.

Titleist alone makes 1 million golf balls a day.

300 million golf balls are lost in the US each year.

Golf balls take 100 to 1,000 years to decompose.

Alan Shepard Jr used a collapsible Wilson Staff Dyna-Power 6 iron to hit 2 golf balls when on the moon. They will never decompose.

Golf is the only sport ever played off planet Earth.

Arrows from a longbow are slow because of their weight at a maximum of 175 fps.

Recurve bow arrow can get to 150 mph or 225 fps.

Compound bow arrows can travel up to 200 mph or 300 fps.

Crossbow arrows 280 to 350 fps, top models at 266 mph or 400 fps or more.

150 mph @ 225 fps = 241 kph @ 68.5 mps.

200 mph @ 300 fps = 321.8 kph @ 91.4 mps.

Mph = miles per hour. Fps = feet per second. Kph = kilometers per hour. Mps = meters per second.

Fastest fish: Sailfish; 68 mph, 109 kph.

Largest animal ever to live on Earth: Blue Whale.

80 to 100 feet long and weigh 441,000 lbs.

Their tongue weighs as much as an elephant

Blue whale is the loudest animal on the planet

Sounds from a Blue Whale can be heard up to 1,000 miles away

They live 80 to 90 years

Their heart weighs about 400 lbs

Blue whale penis is the longest on the planet, over 8 feet long.

The largest testicles belong to the Northern Right Whale which can be 900 kg or 1984 lbs.

Fastest horse. Triple crown winner Secretariat at 37.8 mph. 60.8 kph, in 1973.

Fastest horse ever clocked, Winning Brew at 43.7 mph, 70.3 kph, but Secretariat would beat Winning Brew in the Kentucky Derby's 1¼ mile race.

Fastest dog. Greyhound, 45 mph, 72 kph.

Fastest domestic cat. Egyptian Mau, 30 mph, 48 kph.

Fastest human: Usain Bolt, 27.78 mph, 44.72 kph, 7 Feb 2020.

Fastest bug: Dragonfly, 35 miles per hour or 56.3 kph.

The dragonfly is also the best bug for humans. A dragonfly will eat up to 100 mosquitoes a day, caught in flight. Also, other flies and black flies.

Third fastest insect is the common housefly. The buzzing sound is the fly's wings which beat at up to 20,000 beats per minute. A fly can react to danger and fly in .039 seconds, over twice as fast as the blink of a human eye.

The best and fastest way to catch mosquitoes and flies is a vacuum cleaner. Insects breathe through their skin, and the dust in the vacuum cleaner will suffocate them.

Zappers are also excellent; they cook the bugs. (Sometimes they pop like tiny firecrackers) I always used to carry one when I travelled, great for flies, mosquitoes, spiders, or any bug you can reach. (Be careful, they are zappers, they hurt when touched but will not snap on skin covered with any cloth.)

Fastest tides: Saltstraumen, Norway. 25 mph, 40 kph maximum.

Highest tides, Burntcoat Head, Bay of Fundy, 53.6 ft, 16.3 m.

This causes Reversing Falls, Saint John, New Brunswick, at the head of the Bay of Fundy.

Standard parachute descent rate is 17 mph with a glide ratio of 1:1. (one foot down and 1 foot forward)

HALO is High Altitude Low Opening (Used by military & sport jumpers)

Hang gliders fly between 20 and 30 mph with a glide ratio of 16:1, but can reach speeds of 80 mph

A wingsuit can go as low as 25 mph or as fast as 220 mph.

Wingsuit glide ratio can be 3:1, 3 feet forward to 1 foot down.

1 atmosphere is 14.7 lbs of air per square inch at sea level.

The Karman Line where air pressure is zero is 62 miles (100 km) of altitude.

Blood pressure would blow the human body apart, hence the need for pressure suits at high altitudes above 63,000 feet.

By law, pilots must have oxygen available, O^2, above 13,000 ft.

Blood pressure is measured in millimeters of mercury (mmHg).

Systolic (upper) and diastolic (lower).

33 feet of water equals 1 atmosphere

The Mariana Trench is 11 km or 6.835 miles (36,200 ft) below the sea surface and pressure is 1.1 kbar or 8 tons per square inch. 1,100 times greater than on the earth's surface.

Fastest snake, sidewinder at 18 mph, 29 kph, found in North America.

Black Mamba in Africa 12.5 mph, 20 kph.

Fastest turtle: leatherback, also the largest. 22 mph or 35 kph.

Fastest swimmer:

Male: Michael Phelps, 6'4" tall, 4.71 mph with the most medals in Olympic history with 28. 23 gold, 3 silver, 2 bronze. Net worth $100 million.

Female: Katie Ledecky, 6' tall, 3.78 mph, with 14 Olympic medals. 9 gold, 4 silver, 1 bronze. Net worth $5 million.

The average walking speed of a human being is 2.5 to 4 mph.

Snapping your fingers is 20 times faster than the blink of an eye, which is 1/10th of a second.

The snapping sound is the finger hitting the palm of your hand. The thumb provides the resistance to provide the necessary speed for your finger to hit your palm.

When you are sitting immobile in a chair at home, are you moving?

Your heart is moving blood 3 to 4 miles per hour, or 3 feet per second.

Your eyes are receiving signals at the speed of light and are sending messages to the brain in 13 milliseconds (1 millisecond is 1/000ᵗʰ of a second) or from 179 to 268 mph.

Your body is rotating on planet Earth at 1,000 mph (1609 kph), which is circling the sun at an orbital speed of 67,000 mph (107,826 kph). Our solar system is moving at about 450,000 mph (720,000 kph) around the Milky Way in a universe that is expanding faster than the speed of light. Yes, you are really moving fast, sitting still.

Speed of nerve impulses: 268 mph or 431 kph.

Speed of blood: 3 feet per second or 1 meter per second. A single blood cell travels through the ENTIRE body in about one minute. The heart pumps about 83 gallons per hour. The average body has 60,000 miles or 96,560.64 km of blood vessels, enough to circle Earth more than twice. The average body has about 6 liters or 6 quarts of blood and travels 12,000 miles or 19,000 km per day. Red blood cells live for about 120 days.

Blood moves at 3 to 4 mph. Your blood pressure would explode your body at zero air pressure. The heart also has to pump the blood through the lungs. The surface area of a normal man's lungs is about 100 square meters or 1080 square feet and is called the bronchial tree, made up of 480 million alveoli. The lungs do 2 things: absorb oxygen, O_2 and expel carbon dioxide, CO_2.

Blink of an eye = $1/10^{th}$ of a second.

3 reasons:

-reflex for a foreign object,

-bright light = optical reflex,

-sounds over 40 to 60 dB (decibels),

Contact lenses can diminish the reflex action.

Visual reaction time = 200 – 250 ms (milliseconds),

Hearing = 150 – 200 ms,

Touch = 130 -170 ms.

Normal resting heartbeat is 60 to 100 beats per minute (bpm).

Resting bpm decreases to below 50 bpm in athletes

200 bpm is high for strenuous exercise.

300 bpm is the absolute maximum for the human heart and is very dangerous.

Speed of a cough can be up to 50 mph and expel up to 3,000 droplets of germs.

A sneeze can be up to 100 mph, and up to 100,000 droplets of germs can be expelled from 5 to 30 feet away.

It is very difficult to hold your eyes open when you sneeze because closing the eyes is a reflex action.

It is not possible to sneeze during REM sleep. (Rapid Eye Movement)

During REM sleep, the body is fully relaxed and cannot move.

You can fart or burp when asleep. A Fizzle is a silent fart.

A Furp is a fart and a burp at the same time.

Speed of a human fart is about 10 feet per second or 6.8 mph.

A small volume fart would be the amount of air in a bottle of nail polish, whereas a very large fart would be the volume of air in a soda can. (355 ml or 12 oz)

How fast is 8 seconds?

Ask a cowboy bronco or bull rider. 8 seconds is a long time.

Speed of death is about 4 minutes for permanent brain damage and death 4 to 6 minutes later. The heart lasts the longest.

In deep water a Tsunami can run at up to 500 mph and only a few feet high on the surface.

They slow to 20 to 30 mph when the tsunami reaches shallow water.

The highest tsunami was in Alaska in 1958 at 1,722 feet or 525 meters.

On March 11, 2011, the tsunami in Japan was 132 feet or 40 meters high.

The normal maximum height is up to 98 ft or 30 meters.

Surfers riding a massive wave can achieve speeds of 40 to 50 miles per hour.

Waves larger than 7 feet, the surfer can get up to 20 mph.

Waves 4 to 7 feet, the surfer can get to 10 to 15 mph.

Large waves move at about 30 mph or 50 kph about every 14 seconds.

Speed of urine is about 3.4 mph.

Flow for men is 12 ml/sec or 0.40 oz/sec, through a long narrow channel also designed to spurt semen.

Flow for women is 18 ml/sec or 0.69 oz/sec, through a short wide channel used for nothing else.

Obviously, women can empty their bladder much faster than men.

How fast does a human grow hair and nails?

Human hair grows: .014 in, .35 per day or 5 in, 12.75 cm per year.

2.45 mm or 0.1 in per week.

12.5 mm or 5 in per year.

Fastest hair growth is 15 to 30 years old. Slows after 50 years.

Hair lasts 2 to 7 years and then is replaced.

Male pattern baldness normally starts in their 30's.

Human fingernails grow: 1/8 in or 3 mm/month.

Human toenails grow: .041 in, or 1/16in, 1.62 mm per month.

Slowest mammal:

Sloth, 0.16 mph, 0.27 kph.

3-toed sloth, 0.11 mph, 0.017 kph

The English built the first sailing ship armed with cannon in 1414.

How fast does it take the cold to freeze the balls off a brass monkey?

The brass monkey is the brass triangle that used to hold the iron cannon balls in a pyramid beside a sailing ship's cannon. The brass prevented any possibility of sparks between the iron cannonballs, gunpowder, and the brass monkey. It also kept the cannon balls from sitting on the wooden deck that was often awash in water. The brass contracts faster and more than the iron that the cannon balls were made of and in extreme cold the contraction of the brass monkey would permit some of the iron cannon balls to fall from the pyramid formed by the brass monkey. Hence the expression: "Cold enough to freeze the balls off a brass monkey".

Tectonic plates. There are 7 major and 5 minor plates. The largest and fastest moving is the Pacific plate, which goes from Chili, then up the South American west coast to Mexico, then California (eg: San Andreas fault), then up to Alaska, west across to Kamchatka Peninsula, Russia, then down to Japan and then Indonesia.

Known as the ring of fire.

This plate moves about the same speed as toenails grow, 0.41" (1/16") 1.62 mm per month.

Latest strongest earthquakes:

Kamchatka, Russia in 1952, 9.0.

Chile in 1960, 9.5.

Alaska in 1964, 9.2

Sumatra in 2004, 9.1

Tohoku in 2011, 9.1, lasted 6 minutes and wrecked 3 nuclear reactors in Japan.

The pyramids of Giza have moved 96.75 meters, or 317 feet 5 inches, in 4,500 years since they were built. The pyramids are on the African tectonic plate and are oriented perfectly to the north, south, east, and west.

All of the strongest earthquakes happened at the edges of the Pacific plate, the Ring of Fire. This includes the biggest explosion, which was at Mt St Helens, which was 9,677 ft high and is now only 8,364 feet high, on 18 May 1980. The Mt St Helens explosion was equal to a 400-megaton nuclear explosion, 8 times bigger than the most powerful nuclear explosion ever detonated on earth. Nagasaki was only 21 kilotons, 40% greater than Hiroshima. Mt St Helens was heard 140 miles (225 km) away for 44 seconds. The initial explosion was heard 217 miles (350 km) away.

4 types of earthquakes. Tectonic, volcanic, collapse, explosion.

END

One thing I regret in my life is that I have never taken a catapult launch from the cockpit of a jet aircraft on a US aircraft carrier. That is speed, fast!

A message to Tom Cruise: Everything in Flying Sixth Sense happened. No producer, no director, no script, no stuntman. (although I have heard that you seldom use a stunt man) Reality!

To all that read this, if you have any comments or questions, please contact me at:

flyingsixthsense@yahoo.com.